PLEASE REFER TO PRINTED
RE JE
G000147496

MANAGING WATER FOR AUSTRALIA

Managing Water for Australia

The Social and Institutional Challenges

Editors: Karen Hussey and Stephen Dovers

© CSIRO 2007

All rights reserved. Except under the conditions described in the *Australian Copyright Act* 1968 and subsequent amendments, no part of this publication may be reproduced, stored in a retrieval system or transmitted in any form or by any means, electronic, mechanical, photocopying, recording, duplicating or otherwise, without the prior permission of the copyright owner. Contact **CSIRO** PUBLISHING for all permission requests.

National Library of Australia Cataloguing-in-Publication entry:

Managing water for Australia : the social and institutional challenges.

Bibliography.
Includes index.
ISBN 9 78064309 3928.

1. Water-supply – Government policy – Australia. 2. Water-supply – Australia – Management. 3. Water-supply – Social aspects – Australia. 4. Water resources development – Australia. 5. Water use – Australia. I. Hussey, Karen. II. Dovers, Stephen.

333.910994

Published by:
CSIRO PUBLISHING
150 Oxford Street (PO Box 1139)
Collingwood VIC 3066
Australia

Telephone: +61 3 9662 7666
Local call: 1300 788 000 (Australia only)
Fax: +61 3 9662 7555
Email: publishing.sales@csiro.au
Website: www.publish.csiro.au

Front cover: Lake Grace, Adrianne Yzerman. Overall winner, photography competition, Land & Water Australia, 2006.
Set in 10pt/12pt Minion
Edited by Adrienne de Kretser
Cover and text design by James Kelly
Typeset by Palmer Higgs
Printed in Australia by Ligare

The chapters in this book were produced through a writing and conference process organised and supported by Land & Water Australia, the National Water Commission, The Australian National University and the Academy of the Social Sciences in Australia. The opinions expressed in the book are those of the relevant authors.

Foreword

Current water use in Australia is unsustainable in many areas. The challenges we face to manage our water resources on a truly sustainable basis are formidable. It is important not only to get the science right, but to ensure that the Australian community understands the aims and the processes to achieve sustainable water management so it can support and participate in making the difficult decisions about water management that face us as a nation.

To help meet this need, the conference Delivering the National Water Initiative: Understanding the social and industry dimensions, was hosted by Land & Water Australia and the National Water Commission together with The Australian National University and Academy of Social Sciences in Australia on 4–5 December 2006 at Parliament House, Canberra.

The National Water Initiative is the blueprint for implementing the most significant reforms to the management of Australian freshwater resources for more than a century, and the conference offered the opportunity to explore the challenges that brings. We are pleased to present this book to help build further links between research knowledge and the practical implementation of water reform.

This book brings together the 10 expert papers presented at the conference as a summary of existing social sciences research and knowledge and to stimulate deeper thinking about the implementation needs of the agreed water reform agenda of the National Water Initiative. Key areas are identified where further integrated, multi-disciplinary knowledge and analysis across the social sciences is necessary.

The current drought across much of the country, our acceptance of the challenges that climate change brings, and recognition of the over-allocation of Australia's water resources, highlights the timeliness of these papers on implementation challenges and opportunities.

A national $10 billion plan to accelerate water reform and significantly improve water management was announced by the Australian Government in January 2007. The new plan is in large part underpinned by, and will significantly advance, the National Water Initiative's reforms. This book is now more pertinent than ever as it helps to bridge the gap between the words of the National Water Initiative and the actions needed to give effect to those words on the ground across nine jurisdictions.

Ken Matthews AO
Chief Executive Officer, National Water Commission

Dr Michael Robinson
Executive Director, Land & Water Australia

Contents

Contributors

Harry Abrahams is Team Leader of the Water Reform Group in the National Water Commission. Prior to joining the Commission in 2005, Mr Abrahams was a member of the Australian Government Joint Natural Resource Management Team, working on the delivery of the National Action Plan for Salinity and Water Quality and the Natural Heritage Trust in both Queensland and the Northern Territory.

Kathleen Bowmer is Professor of Water Policy at Charles Sturt University and Honorary Research Fellow, CSIRO Land and Water. Professor Bowmer has worked extensively in the policy and regulatory sector in natural resource management in Australia for more than 30 years and was a winner of the 1994 Eureka Prize for Environmental Research, awarded by the Australian Museum.

Serena Chen is a PhD student in the Fenner School of Environment and Society at the Australian National University. She has a BSc (hons) from the ANU and is studying ecological function in riverine systems.

Daniel Connell is an environmental historian at the Crawford School of Economics and Government, Australian National University and the author of *Water Politics in the Murray-Darling Basin*. Based on his PhD undertaken at what is now the Fenner School of Environment and Society, Australian National University, his book assesses water management in the MDB against the National Water Initiative. It also tests the adequacy of the NWI against the challenges facing the MDB.

Stephen Dovers has interests in policy and institutional dimensions of sustainable development, and is Professor at the Fenner School of Environment and Society, Australian National University, and Adjunct Principal Research Fellow at Charles Darwin University. His latest book is *Environment and Sustainability Policy* (Federation Press, 2005).

Douglas E Fisher holds the degrees of MA, LLB and PhD from the University of Edinburgh. He is currently Professor of Law at Queensland University of Technology. He has had professional and academic experience in environmental resources law in three jurisdictions over 40 years.

Alex Gardner is a senior lecturer at the University of Western Australia Faculty of Law, a visiting lecturer in the Australian National University College of Law graduate environmental law program, and a senior sessional member of the WA State Administrative Tribunal.

Geraldine Gentle is Director of URS Economics and Policy, and is an economist with extensive experience in microeconomic reform and natural resource management. Previously, she was Assistant Commissioner on the Productivity Commission, where she led many inquiries into complex economic, public policy, regulation and planning issues. She has also been Deputy Director-General of the Queensland Department of Natural Resources and Mines, and Commissioner of the Murray-Darling Basin Commission.

R Quentin Grafton is Research Director of the Crawford School of Economics and Government, Australian National University. He is listed in the Top 500 of the World's Economists (1994–1998), in the 2005 and 2006 Marquis editions *Who's Who in the World*, the Marquis 9th edition *Who's Who in Science and Engineering* and in the 33rd edition of the *Dictionary of International Biography* (in press). He has written over 50 journal articles, book chapters and several important texts in environmental economics, including *The Economics of the Environment and Natural Resources.*

Steve Hatfield-Dodds is an economist and policy analyst with CSIRO Sustainable Ecosystems. He has an extensive career in consulting and was Director of the Socio-Economic Integration Emerging Science initiative for CSIRO.

Karen Hussey is a Postdoctoral Fellow at the Research School of the Humanities, Australian National University, and Chair of the ANU Water Initiative – a transdisciplinary research and education initiative focused on water resource management. Her interests include environmental politics and economics, water policy and management, agri-environment policy (with a particular focus on international approaches) and global environmental governance.

Sue Jackson is a senior research scientist with CSIRO's Division of Sustainable Ecosystems and an executive member of the Tropical Rivers and Coastal Knowledge (TRACK) Research Hub. Her current research focus is Indigenous values associated with water and Indigenous participation in water resource management frameworks.

Tony Jakeman is Professor at the Fenner School of Environment and Society and Director of the Integrated Catchment Assessment and Management Centre, Australian National University, and has been an environmental modeller for 30 years with over 300 publications in the open literature. His interests include integrated assessment methods for water and associated land resource problems, as well as modelling of water supply and quality problems including in ungauged catchments.

Rebecca Letcher is a Fellow at the Australian National University and has contributed over 50 journal publications to the water resources literature. Her research interests relate to the development of approaches for integrated assessment of natural resource management issues. This includes development of integrated modelling approaches and frameworks, development and application of participatory approaches in modelling, and the design of decision support systems.

Joe Morrison is the executive officer of the North Australian Indigenous Land & Sea Management Alliance (NAILSMA) and an executive member of the Tropical Rivers and Coastal Knowledge (TRACK) Research Hub. NAILSMA has recently established an Indigenous Water Policy Group to increase Indigenous participation in northern water policy debates.

Chris Olszak is a Senior Economist at URS Australia, and has led numerous evaluations of Australian water projects. They include reviews of water allocation, infrastructure and system management, recycling, water-saving initiatives, environmental flows and other environmental actions. He recently advised the NSW Natural Resource Commission on reviewing water-sharing plans. In 2005, he worked in the Department of Water Resources Management in Vietnam where he led a USAID-funded investigation into the development of water allocation systems and the use of economic tools.

Deborah Peterson is an Assistant Commissioner at the Productivity Commission, heading the Environmental and Resource Economics Branch. Previously, she has held senior positions at the Australian Bureau of Agriculture and Resource Economics and at the Bureau of Industry Economics, and was President of the Australian Agricultural and Resource Economics Society in 2005. She is currently a Visiting Fellow at the Ecole Nationale Supérieure d'Agronomie de Montpellier in France.

Lisa Robins is a PhD student at the Centre for Resource and Environmental Studies, Australian National University, researching capacity-building measures for regional natural resource management. Principal of Robins Consulting since 1998, she has 17 years' experience in natural resource management, including coordinating the Dryland Regions Strategic Investigations & Education Program for the Murray-Darling Basin Commission (1998–2004) and leading the Catchment Team for the National Dryland Salinity Program (2003–2004). She was awarded the 2004–2005 Environmental Change Institute-Green College Teaching Fellowship, at Oxford University.

Kate Stoeckel is a Policy Officer in the Water Reform Group in the National Water Commission. Before joining the Commission in 2006, she was a legal advisor to an NGO in Cambodia on land reform and practised as a corporate lawyer in Canada and Hong Kong. She has economics and law degrees from the University of Sydney, and has undertaken a Masters in Laws specialising in natural resources and water law at the University of British Columbia, Canada.

Geoff Syme is Research Director, Society, Economy and Policy, CSIRO Land and Water. He has been researching social aspects of water policy for 30 years.

Mike Young holds a Research Chair in Water Economics and Management at Adelaide University and Adjunct Professorships at the University of New England and Charles Sturt University. A Fellow of the Academy of the Social Sciences in Australia, until recently he was Chief Research Scientist with CSIRO Land and Water. In 2003, he was awarded a centenary medal for outstanding service through environmental economics, and in 2005 he (with Jim McColl) was awarded the Land and Water Australia Eureka Prize for Water Research.

Introduction
Informing Australian water policy

When the well is dry, we know the Worth of Water.
(Benjamin Franklin, *Poor Richard's almanac*, 1745: 2)

In recent years, water planners have recognised a simple three-part sequence to characterise the developmental phases of water resources. The first phase applies to pre-industrial societies where water is regarded as a free gift and easily accessible. The second phase is distinguished by active water exploitation, the construction of dams for hydro-power and irrigation, and inter-basin transfers from better endowed regions to nearby dry regions. The final, mature phase has close to maximum attainable level of stream flow regulation in major river basins, the costs of further water resource development and management increase rapidly and attention turns to non-conventional techniques to enhance supply (Smith 2003: 53). For Australia, owing to our small population base and relatively short history of settlement, the arrival and recognition of the mature phase has been much more recent than for most countries. But the 1980s heralded a sea-change in water resources planning so that changes in water policy over the last two decades have, arguably, eclipsed that of the preceding 90 years.

The most recent development in Australia's water planning has been the launch of the National Water Initiative (NWI) in 2004, with a schedule of implementation to 2014. Agreed to by the Commonwealth and all state and territory governments, the NWI is the overarching policy framework guiding Australian water management. It reflects and significantly extends key policy reforms in Australia over the past two decades, and brings these together into one powerful agenda which incorporates, among other things, integrated catchment management, tradable water rights, full accounting of resources and use, regional water planning, and environmental allocations (see Chapter 1, this volume, for a detailed description). In this respect, the NWI, and the key elements and principles therein, reflect the modern idea of sustainability and Australia's commitment to ecologically sustainable development (ESD). ESD is not about environment, per se, but about an integrated policy agenda that is wider and deeper, incorporating the long term integration of social, ecological and economic imperatives, more precautionary approaches to the environment, including people in policy and management, and creating new institutions and processes.

Through the NWI, Australian government and major non-government interests have established agreed directions for water policy and management over coming decades. As a decade-long national policy framework, the NWI allows a longer term outlook than most other policy initiatives, and thus the ability to work through many remaining tensions and potential implementation difficulties. This is a significant opportunity, as the reform agenda the NWI sets out is indeed an ambitious and difficult one, and the magnitude of the task is only now beginning to be realised. Assumptions regarding implementation are being unsettled by realisations of significant deficits of capacity and knowledge. Having agreed to the policies outlined in the NWI, can those policies be achieved based on existing knowledge and institutional capacities? This book addresses the major challenges in implementing the NWI with particular focus on social sciences research and knowledge that can and should inform policy and decision making.

Common themes

The chapters in this book are the culmination of an ongoing process of development and negotiation comprising two main stages. First, Land & Water Australia (LWA) established a social and institutional research agenda linked to implementation challenges defined in the NWI. This research agenda was subsequently published in summary format in *Water perspectives* (LWA 2005). Second, agreement was reached between LWA, the National Water Commission, The Australian National University and the Academy of the Social Sciences in Australia to pursue this agenda in greater detail and rigour via the commissioning of papers by leading Australian researchers, which were designed to bring together existing relevant knowledge and identify gaps. The range of disciplines covered include anthropology, history, sociology, government and policy areas, institutional and public administration, and cultural, economic and legal areas.

The participating researchers addressed ten research and knowledge areas that match policy implementation areas within the NWI, as well as some common themes and topics to guide analysis along comparable lines. Each chapter is written to a standardised format addressing the following:

- the nature of the policy implementation task, for example: creating and managing water markets, establishing comprehensive water plans, governing environmental flow allocations, etc.
- what we already know, or what knowledge we can quickly adapt, to support policy formulation, implementation and evaluation.
- what critical knowledge needs exist that will require targeted research and knowledge generation/transfer, to enable achievement of agreed national water policy goals fulfilling ecological, social and economic goals.

Interaction during the course of this project ensured dialogue between individual researchers and review of emerging analysis and conclusions. Two authors' workshops during the project provided opportunity for discussion prior to a two-day national conference in Parliament House, Canberra, which provided yet more opportunity for debate and distillation of the key challenges in implementing the NWI and, crucially, of how the social sciences might contribute. The book has a strong and constructive focus on the reform agenda for water policy, as enunciated in the NWI, but also considered in the context of previous, less coordinated reform.

It is evident that the challenges in implementing the NWI are many and exacerbated by Australia's federal system of government, the different systems of management in each of the States and Territories, the different sets of legal arrangements in each of the States and Territories and the limited capacity of the Commonwealth to deal with the management of water resources. The NWI contemplates a set of cross-jurisdictional related planning, regulatory and market arrangements but the juxtaposition of these three functions poses particular challenges for an integrated and coherent set of policy and legal arrangements. The NWI's focus on a nationally consistent approach to water planning, underpinned by the objective of water trading, will require a consistent and coherent legal framework across water resources as a whole in ways that are intrinsically enforceable as a matter of law. The need for a strong legal framework is similarly highlighted in relation to environmental water allocations.

A second theme emerging from the chapters is the need for more (and better) communication in the reform process and greater public involvement in policy and planning frameworks. The objectives of the NWI are ambitious, in many cases contentious, and not yet clearly understood by many people, so there is a very real need to bring the community along. Experience from reforms across a range of policy sectors, especially where resource rights are being revised, strongly suggest that the successful implementation of the NWI will depend on the communication of its underlying logic, objectives, anticipated impacts and decision-making processes to key stakeholders, whether they be from industry or the community (Connor and Dovers 2004).

A related challenge is the need for greater research and training in the development of water plans and capacity building amongst water planners. As the primary conduit for the implementation of the NWI is regional and local government agencies, those agencies must in turn be provided with the necessary detail and know-how, including: design and implementation of planning and review processes, appropriate and effective participation; practical guidance on analysing trade-offs in water allocation and management, and how to manage the inevitable social issues arising from those trade-offs; and information and training on how to integrate better the objectives of water planning with other natural resource management goals. The role of the social sciences in providing that knowledge is manifest.

We should also note that, in almost all the chapters, a lack of scientific knowledge and data was highlighted as a substantial impediment to the successful implementation of the NWI. Perhaps a similar exercise focusing on knowledge gaps in the natural sciences might 'complete the picture' and facilitate further the implementation of agreed national water reforms.

As this book was going to press, intense discussion was taking place around Commonwealth proposals to take control of water policy responsibilities in the Murray-Darling Basin. However this issue of responsibility within the federal system resolves, the reform implementation tasks dealt with in this book remain. The chapters within offer a substantial, rigorous and highly topical contribution of knowledge to Australia's water resource management capacities.

Karen Hussey and Stephen Dovers

Chapter 1

Water reform in Australia

The National Water Initiative and the role of the National Water Commission

Kate Stoeckel and Harry Abrahams

Water availability and management is a key issue for Australia. It is also now highly topical. Limited water resources and the disparate – and potentially conflicting – economic, environmental and social interests served by water mean that effective water planning is an absolute necessity. The variability of our climate and the location of water resources compared with growing population centres, and the importance of rural industries reliant on irrigation, create a suite of circumstances particular to Australia. No-one can overlook the impact of the recent and frequent drought conditions in Australia complicated by (or indicative of) the expected further impacts of climate change. Indeed, two combining factors cause a significant water challenge: first, a significant period of climatic drying has been observed; and second, the majority of systems in southern Australia (the most agriculturally productive area) are fully allocated or overallocated. Water usage, water productivity, water technology and water governance need to be improved to ensure that the needs of a growing urban population and environmentally sustainable and productive river systems are met.

Australia's water consumption has significantly increased in both the urban and rural settings. Most significantly, irrigation consumption has grown exponentially in the last few decades and now accounts for approximately 75% of total water use in Australia (National Land & Water Resources Audit 2001: 56). Figure 1.1 indicates the growth in irrigation areas across Australia from 1920 to 2000.

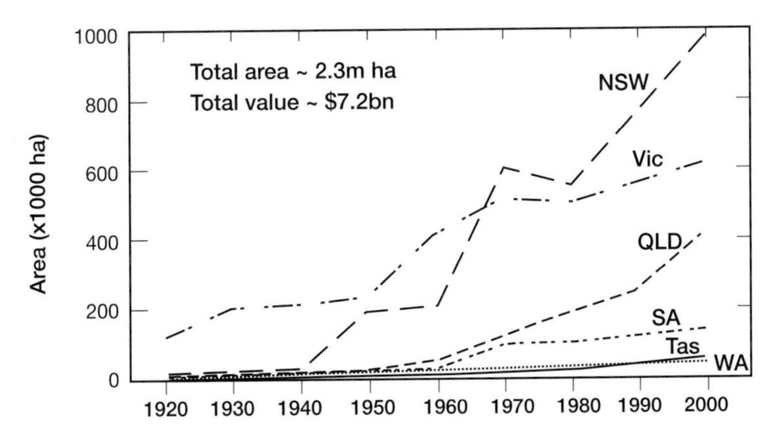

Figure 1.1 Growth in irrigation in Australia, 1920–2000
Data source: Chartres & Williams (2006).

Water issues have been recognised as vital to Australia since Federation. Indeed, the future of the Murray River proved a contentious issue for the colonies of New South Wales, Victoria and South Australia during Federation discussions, particularly against the backdrop of the Federation drought that lasted from the mid 1890s to 1902 (La Nauze 1972). Constitutionally, responsibility for resource planning and management rests with each state and territory; they then allow other parties the rights to access and use water for a variety of purposes, for example irrigation, industry, mining, recreation, and servicing rural and urban communities.

Water does not respect state boundaries. Geography and connectivity between water systems dictate that better water management and planning are, in fact, national concerns that require national action. A nationally consistent approach is also needed to achieve the necessary requirements for market systems and trading of water – key elements underpinning current water reform efforts – such as enhancement of resource investment security, adequate pricing, market transparency and good water accounting.

We are at a crucial stage in implementing water reform in Australia. Recognising the need for a national framework for water reform, the Council of Australian Governments (COAG) agreed to a strategic framework for the efficient and sustainable reform of the Australian water industry in 1994 (the 1994 COAG Framework).[1] Significant progress towards more efficient and sustainable water management was achieved under this agenda, but the need for national water reform persisted despite the changes and commitments.

Demand for water has increased since 1994 but progress with water reforms between regions and jurisdictions has remained variable, placing even greater pressure on water reform efforts. The knowledge base regarding surface and groundwater systems and the requirements for effective and efficient water markets expanded concurrently, creating an opportunity to complement and extend the reform agenda. This continued imperative for water reform at the national level resulted in all governments agreeing to a modified and extended reform agenda in 2004 designed to more fully realise and extend the benefits intended by the initial 1994 COAG framework. This reform agenda, formally known as the Intergovernmental Agreement on a National Water Initiative (NWI), is the national blueprint for water reform in Australia and is agreed to by all governments.[2]

The NWI is the outcome of a detailed and negotiated process. It represents a clear way forward but recognises that there are significant uncertainties, knowledge needs and challenges in implementing those commitments. To fully implement the initiative, we need new research and knowledge management approaches in addition to the identified administrative, institutional and policy reforms outlined in the NWI itself. Addressing these complex and interdependent research questions requires a highly collaborative and participative approach.

Often, environmental and productive issues take the limelight in the reform debate, but there is a clear and growing recognition of the importance of the social science agenda required to implement effective water reform. However, the social and industry ramifications of water reform actions – and inaction – are currently little understood. To progress the social science research agenda linked to the implementation of the NWI, Land & Water Australia's (LWA) Social and Institutional Research Program facilitated a workshop in May 2005 to look at social and institutional research questions in support of the implementation of the NWI. A number of social science experts with water-specific expertise, from a range of disciplines, discussed opportunities for research across ten themes they identified as the main areas of social and institutional implementation challenge to the NWI (discussed below). Workshop findings were published in the report *Water Perspectives* (Land & Water Australia 2005).

These ten social and industry dimensions of water reform are explored in an expert paper on each topic. The papers are designed to provide a compendium of information and establish a constructive and critical link between social science research and water policy in Australia, while outlining how further research can contribute to implementing water reforms. Recognising the importance of a cross-disciplinary approach and aiming at adaptable application, the publication focuses on the current status of knowledge and capacity of the social sciences, and where gaps in the research are most prominent.

1 A blueprint for national water reform: the NWI

Since the mid 1990s Australia has witnessed a great change in the approach to reform of water policy and management under the 1994 COAG framework. States and territories have made considerable progress towards more efficient and sustainable water management. Most jurisdictions have embarked on significant reform programs of their water management regimes, separating water access entitlements from land titles, separating the functions of water delivery from those of regulation, and making explicit provision for environmental water.

However, a lot more still needs to be done to meet the challenges ahead. Significant gaps in knowledge remain, particularly regarding groundwater, environmental and ecological needs, and deficiencies in measurement, accounting, metering and reporting. Water planning and water policy-making are immature, and water markets are underdeveloped even though the 1994 commitments were to address these issues as matters of priority. These are all difficult issues and represent major reforms to the way we manage water. The Australian Water Resources 2005 report confirms the need for more and sustained action, noting considerable room for improvement in water resource planning, water resource development and management of river and wetland health (National Water Commission 2006a). When placed against the backdrop of one of the worst seasonal outlooks with respect to water in Australia's history (National Water Commission 2006b), water reform remains a key issue facing Australia.

The continued need for water reform at the national level to extend previous COAG commitments resulted in the NWI, a landmark document, keeping water at the forefront of national public policy. It provides a framework to address and deliver the more difficult COAG water reform commitments where little progress had previously been made,[3] as well as instigating new commitments. The NWI clarifies the key actions to be undertaken and enunciates a clear pathway to achieve a national integrated water reform agenda with broad parameters for water policy and management from 2004 to 2014, and encompasses many different players and issues.

As well as extending the 1994 COAG framework, the NWI was designed to complement other natural resource management initiatives with a significant water focus that are the subject of separate agreements by the parties to the NWI, particularly the National Action Plan for Salinity and Water Quality and the Natural Heritage Trust. Continued implementation of the National Water Quality Management Strategy also complements NWI outcomes. Recognising the need and desirability for a nationally consistent approach, governments have agreed that the NWI will take precedence over the other water management agreements to the extent that any inconsistencies arise (NWI Preamble, para. 7).

The overall objective of the NWI is to increase the productivity and efficiency of Australia's rural and urban water use while ensuring community needs are met and river and groundwater systems are returned to environmentally sustainable levels of extraction (NWI Preamble, para. 5). We need greater certainty for investment and the environment, both to meet these aims and to underpin the capacity of Australia's water management regimes to deal with change responsively and fairly. The NWI is a firm and positive commitment by all governments in this direction.

Several factors are critical to the success of the NWI, including a wide range of governance reforms, definition and determination of environmentally sustainable levels, definition of water budgets that indicate volumes of water available for use and trade under varying climatic conditions, and availability of measurement and monitoring technologies to underpin robust accounting systems. The NWI states that its overall goal of optimising economic, social and environmental outcomes through a nationally compatible market, regulatory and planning-based system of surface and groundwater resource management will be met by achieving the following ten objectives (NWI, para. 23):

1. Clear and nationally compatible characteristics for secure water access entitlements.
2. Transparent and statutory-based water planning.
3. Statutory provision for environmental and other public benefit outcomes, and improved environmental management practices.
4. Complete the return of all currently overallocated or overused systems to environmentally sustainable levels of extraction.
5. Progressive removal of barriers to trade in water, and meeting other requirements to facilitate the broadening and deepening of the water market, aiming for an open trading market.
6. Clarity of the assignment of risk arising from future changes in the availability of water for the consumptive pool.
7. Water accounting which can meet the information needs of different water systems in respect to planning, monitoring, trading, environmental management and on-farm management.
8. Policy settings which facilitate water use efficiency and innovation in urban and rural areas.
9. Addressing future adjustment issues that may affect water users and communities.
10. Recognition of the connectivity between surface and groundwater resources and connected systems managed as a single resource.

To achieve these objectives the NWI clearly outlines eight areas where specific work is required (para. 24):

1. **Water access entitlements and planning framework**, separating water from land and focusing both on providing protection for water entitlements (including for environmental water) by enshrining them in statute, and on providing an adaptive framework for water management.
2. **Water markets and trading**, to facilitate the operation of efficient water markets and opportunities for trading within and between jurisdictions.
3. **Best practice water pricing and institutional arrangements**, promoting economically efficient and sustainable use of water resources, infrastructure and government water management resources, while facilitating the efficient functioning of water markets under the principles of user-pays, transparency, and cost recovery.
4. **Integrated management of water for environmental and other public benefit outcomes**, where the agreed outcome is to identify within water resource planning frameworks, the environmental and other public benefit outcomes sought for water systems, and to develop and implement management practices and institutional arrangements that will achieve those outcomes.
5. **Water resource accounting**, whereby parties have agreed to ensure that adequate measurement, monitoring and reporting systems are in place in all jurisdictions, and to support public and investor confidence in the amount of water being traded, extracted for consumptive use, and recovered and managed for environmental and other public benefit outcomes.

6. **Urban water reform**, focusing on the provision of safe and reliable water supplies, increasing water efficiency in urban settings, and encouraging reuse and recycling, as well as innovation in water supply. This element also links with the improved pricing and the water trading outcomes.
7. **Knowledge and capacity-building**, recognising the ongoing priority for increased knowledge and capacity-building across a number of areas in order to support implementation of the NWI.
8. **Community partnerships and adjustment**, under which the agreed outcome is to engage water users and other stakeholders in achieving the objectives of the NWI by building confidence in the reform process, ensuring transparency in decision-making, and providing the public with sound information at key decision points.

A range of specific policy and management tasks are established under each of these eight NWI elements for public, private and community interests in the water sector – a blueprint for action. Each element clearly identifies actions for implementation, including state-specific actions, group actions (for example, interstate water trading commitments to be met by Southern Murray-Darling Basin states), a number of coordinated national actions, and National Water Commission obligations. Every state and territory is required to submit an implementation plan for accreditation by the National Water Commission that details how it will address NWI actions in the coming years. Nearly half of the NWI's approximately 70 specific actions involve national actions or other action by governments working together. This reflects not just its emphasis on greater national compatibility in the way Australia measures, plans for, prices and trades water, but also represents a greater level of cooperation between governments to achieve this end.

Ten key areas of opportunity were identified at the LWA *Water Perspectives* workshop where social and institutional research and synthesis can support the implementation of the NWI. In some cases these correspond directly to the ten objectives and eight key elements of the NWI, but the scope of, and interrelation between, many social science disciplines means that the set of key implementation areas identified were slightly different.

1. **Integrated assessment of the impacts of policy and water allocation changes across social, economic and environmental dimensions**, in particular examination of the suitability of existing and new methodologies to facilitate more confident interrogation of research questions and determination of reliable answers to avoid conflicting attempts to solve water resource issues.
2. **Water plans and accreditation in regard to content requirements and processes**, recognising that successful implementation of the NWI demands more innovative planning and an adaptive management approach to respond to rapidly changing circumstances in the environment, in knowledge and skill, and in community and stakeholder interests.
3. **Linkages between rural and urban water systems, including in peri-urban areas**, identifying consequences for rural systems of supply shortfalls in urban and peri-urban systems and vice versa by examining determinations of the value of water to the economy, pricing and demand management issues, and policy impediments to linking rural and urban policies and systems.
4. **Indigenous perspectives in water management, reforms and implementation**, examining in detail the implementation areas of the NWI relating to Indigenous interests and defining and understanding Indigenous water values, rights, responsibilities and use.
5. **New frameworks for law and regulation, and current settings as enablers or constraints on reform implementation**, considering the requirements for a robust

legal and regulatory framework that will support the implementation of the NWI while meeting community and investor expectations for fairness and transparency.

6. **Values attached to water and their shaping of understanding and communication of reform objectives and implementation**, discussing effective mechanisms for public engagement and integrated decision-making to gain a more sophisticated understanding of how values form and change.
7. **Auditing and review of policy and water plans for effectiveness and appropriate performance measures for impact detection and management**, which examines what data is needed both to inform policy development and to test the value and impact of policies once they are implemented.
8. **Water markets, pricing, trading and transaction costs, and their establishment and functioning**, recognising the important role for nationally consistent water markets and pricing regimes and assessing the ability of these regimes to integrate multiple uses and values of water.
9. **Environmental water allocations and their governance**, which considers the defensibility of environmental flows in legal, cultural and administrative terms, as well as the required elements of effective enforcement and administrative regimes.
10. **Institutional roles, agency responsibilities and capacities in reform implementation**, examining how the NWI's reform agenda changes the roles and responsibilities of water policy or broader natural resource management agencies across Australia and assesses their capacity to manage reform.

The chapters in this book delve into each of these key areas in more detail, discussing each topic's research connections with the social and industry dimensions of implementing the NWI. The book also includes a chapter discussing international policy and management approaches.

2 Implementing the NWI: the role of the National Water Commission

As part of the broader national water reform agenda, the National Water Commission (the Commission) was established to oversee implementation of the NWI and drive the national water reform agenda. The Commission is an independent statutory body in the Environment and Water Resources portfolio, established under the *National Water Commission Act 2004* (Cth) (NWC Act).

The Commission consists of seven Commissioners who are appointed for their expertise in a range of water-related fields (including freshwater ecology, hydrology, resource economics and public sector management) rather than as representatives of sectoral or government interests. The Commission is supported by three groups of staff. The Water Reform Group oversees the implementation of the NWI, develops assessment frameworks, guides states and territories in relation to implementation plans, and monitors and reports on the progress of state and territory agencies in implementing the NWI. The group also manages the Raising National Water Standards Programme. The Water Programs Group manages the Water Smart Australia program, a key program of the Australian Government Water Fund. The Corporate Governance and Strategy Team assists in providing strategic direction to the Commission and provides a range of corporate services to enable other Commission staff to meet their obligations and commitments.

As the key institutional champion for the NWI, the Commission fulfils several main functions in pushing the national water agenda forward (NWC Act, sections 7(1) and 7(2)). The

Commission has a general policy and advisory role, one aspect of which is advising and reporting to the Minister for the Environment and Water Resources and COAG on water issues. Another Commission function in this regard is to assist the states and territories in their delivery of NWI objectives and outcomes aimed at achieving a reformed water business across the country. Indeed, the Commission has a major role to play in encouraging cooperation among all governments towards consistency in water management practices. To this end, the Commission acts as lead facilitator on certain NWI actions such as compatible registers of water entitlements and trades, and nationally consistent approaches to pricing. The Commission is also charged with monitoring the impact of interstate trade in Southern Murray-Darling Basin.

The Commission advises on and administers Commonwealth financial assistance under two programs of the $2 billion Australian Government Water Fund – the $1.6 billion Water Smart Australia programme and the $200 million Raising National Water Standards programme. The Water Smart Australia programme aims to accelerate the development and uptake of smart technologies and practices in water use across Australia, and is targeted at large-scale projects. The Raising National Water Standards programme is designed to help achieve NWI outcomes by investing in Australia's national capacity to measure, monitor and manage its water resources. The Commission is charged with recommending projects for decision by the federal government under both of these programmes. The third program of the Australian Government Water Fund, the $200 million Community Water Grants programme, is administered by the federal Department of the Environment and Heritage jointly with the Department of Agriculture, Fisheries and Forestry.

The Commission also has assessment responsibilities. Initial assessments of Australia's water resources and an assessment of the governance, management and regulation of those resources are underway. Progress assessments of Australian governments in implementing the water reform actions required under COAG 1994 National Competition Policy commitments have been undertaken, with the finding supported by the federal government. The Commission also approves and accredits the state and territory implementation plans. All jurisdictions have submitted plans for consultation, and all of those except Western Australia (due to its recent signing of the NWI) have been accredited.

Every two years (starting in 2006/07) the Commission assesses the performance of the water industry in managing and using Australia's water resources, against national benchmarks. The NWI will be reviewed comprehensively by the Commission in 2010/11, including an assessment of the NWI against performance indicators, an analysis of the extent to which actions taken under the NWI have improved the sustainable management of Australia's water resources, and an assessment of the impact of the implementation of the NWI on regional, rural and urban communities (NWC Act, section 7(2)(i)).

It is important to note, however, that it is not the Commission's sole responsibility to implement the NWI commitments. Primary responsibility lies with the state and territory signatories. Jurisdictions and the Commission have put a number of significant arrangements in place to coordinate implementation of the NWI, especially for actions requiring some degree of national consistency or coordination.

As each jurisdiction has signed the NWI, Ministerial Council processes are being used to implement the actions of the NWI. The Natural Resource Management Ministerial Council (NRMMC) leads most of the joint efforts and the Environment Protection and Heritage Council leads some too. The importance of water in the national agenda led, for the first time, to a water ministers' meeting in June 2006 dedicated to water issues. COAG agreed at its meeting in July 2006 that water ministers should continue to meet as part of the NRMMC.

A number of Standing Committees and policy committees involved with water management exist under the Ministerial Council frameworks. The most significant development has been the creation of the NWI Implementation Committee, which includes officials from the relevant agency of each state and territory government and representatives from the Commission, the Primary Industries Standing Committee and the Environment Protection and Heritage Standing Committee. The NWI Implementation Committee has an agreed work plan that includes an agreed suite of actions to be undertaken in common by all jurisdictions. It meets regularly and provides a coordinated focus for many other working groups and steering committees that relate to specific commitments, such as the Benchmarking Roundtable, the Accounting Expert Advisory Panel, the Performance Indicators Task Group and the Compatible Registers Working Group.

3 Embracing the challenges of water management

Australians can no longer take freshwater resources for granted. Although we live on the driest inhabited continent, we are the fourth-highest water users per capita in the Organisation for Economic Cooperation and Development (OECD 2006: 13). With extremely variable rainfall and common droughts, water scarcity is becoming a way of life. Both rural and urban water users are affected, as well as the environment. Urban water users will face tougher water restrictions; indeed, South Australia has already announced the harshest water restrictions in its history to try to counter record low Murray River flows (Shanahan & Warren 2006: 1). Irrigators will face drastic cuts to their allocations, with some receiving no water at all in the summer 2006–07 season. Scarcity is becoming a major issue and is creating a demand for investigating new sources of water, including desalination and recycling.

Challenges of federation, the differences between state and territory management of water resources and the lack of national frameworks mean that good Commonwealth, state and territory cooperation is essential. However, water can become a very political issue, with strongly held views across all sectors of the community. Detailed research into the social and industry dimensions of water reform may help to depoliticise aspects of the water agenda and thereby facilitate greater cooperation between players and ongoing goodwill towards reform and the NWI in general.

It is important to bear in mind that reform takes time. The strength of the NWI is partly in its long-term outlook. However, sustaining the pace and momentum of reform will require ongoing commitment at all levels of government and sustained community support. There will be many opportunities for stakeholder and community engagement through the reform process, as well as increased opportunity for private sector participation. The social science research agenda can help to inform these debates and consultation processes in order to progress the national reform agenda.

Australia is at a crossroads in terms of its ability to cope with increasing water scarcity, and the NWI is a critical element in moving national water reform in a positive direction. Although a vigorous reform process is underway, many areas still require detailed research. We need greater research on the implications of reform for society and industry and how this affects the implementation of the NWI. The following chapters discuss emerging research and knowledge gaps across a number of key social science issues relevant to successful and sustainable NWI implementation.

A final word

The current water hardships do have an upside, in that both the focus on water reform in Australia and the momentum to drive reform are at an unprecedented level. This renewed commitment to reform provides a strong impetus and support for the implementation of the

NWI reform agenda. The time is right for us to move water reform commitments forward boldly.

The NWI provides us with a wide-ranging, nationally agreed blueprint to progress the reform agenda. Social issues are fundamental to ensuring public ownership of the reform agenda and continued commitment to (sometimes difficult) changes. The importance to reform of the social and industry dimension, the current status of information on those matters and the importance of improving our performance in these areas cannot be underemphasised.

References

Chartres CJ & Williams J (2006). Can Australia overcome its water scarcity problems? *Journal of Sustainable Developments in Agriculture* **1**, 17–24.

La Nauze JA (1972). *The making of the Australian constitution.* Melbourne University Press: Melbourne.

Land & Water Australia (2005). Water perspectives: outcomes of an expert workshop scoping social and institutional research questions in support of the implementation. Available online at www.lwa.gov.au/products.asp.

National Land & Water Resources Audit (2001). Australian Water Resources Assessment 2000.

National Water Commission (2006a). Australian Water Resources 2005: a baseline assessment of water resources for the National Water Initiative, Key Findings of Level 1 Assessment.

National Water Commission (2006b). Australia's water supply status and seasonal outlook October 2006.

National Water Commission (2006c). 2005 National Competition Policy: assessment of water reform progress.

OECD Environmental Performance Reviews (2006). *Water: the experience in OECD countries.* Online at http://www.oecd.org/dataoecd/18/47/36225960.pdf (date accessed 30 Oct. 2006).

Shanahan D & Warren M (2006). States 'fail' on water crisis: cities facing tough new restrictions. *The Australian*, 13 October.

Endnotes

1 The 1994 COAG framework was amended in 1996 to include groundwater and stormwater management revisions and by the tripartite agreement in January 1999 (NWI Preamble, para. 6).

2 The Commonwealth of Australia and the governments of New South Wales, Victoria, Queensland, South Australia, Australian Capital Territory and the Northern Territory signed the NWI at the 25 June 2004 COAG meeting. Tasmania became a signatory to the NWI on 3 June 2005 and Western Australia on 6 April 2006.

3 These commitments are identified by National Competition Policy assessments, which are undertaken to evaluate jurisdictions' performance against water reform framework commitments. They have been and will be undertaken with respect to all COAG water reform commitments, including the 1994 COAG framework and the NWI. The last such assessment was undertaken by the Commission and published in 2006 (National Water Commission 2006c).

Chapter 2

The role of communication and attitudes research in the evolution of effective resource management arrangements

Geoffrey J Syme and Steve Hatfield-Dodds

Water has been a central focus of human culture and society throughout time. Water management arrangements reflect social values, and the ways these values are articulated through communication and institutions. Water values are diverse. A single body of water can provide for survival, wealth, identity and status. Water frequently plays a central role in our appreciation and understanding of our landscapes, providing a basis for fundamental aspects of the human psyche and spirituality (Strang 2004). In an everyday sense, access to water for transport and irrigation has shaped patterns of regional development and been an important driver of many of our formal and informal institutions (Smith 1998; Delli Priscoli 1998).

Changes to water management arrangements, such as the National Water Initiative (NWI), must negotiate these multiple layers of value and meaning and recognise the social, economic and environmental consequences of reform. This adds a social layer to the already formidable technical complexity of water reform. Effective and worthwhile reform requires community support which in turn, needs a sound understanding of the values potentially affecting change, and clear communication of the overall benefits to be achieved by the proposed reforms (Ostrom 2005).

This chapter focuses on the role of communication and community engagement in implementing worthwhile water reform. Section 1 provides a social choice perspective on water management issues as wicked problems, and the various criteria applied in assessing whether a change constitutes 'an improvement'. Sections 2 and 3 concentrate on the essential elements of communication and process we will need to move forward, summarising the Australian literature on communication issues and on attitudes and behavioural change relevant to water reform. Section 4 discusses the role of social values and public involvement in policy and planning and the potential to better engage communities in improving water outcomes. The chapter concludes that more systematic attention to community attitudes and values, backed by the right social and economic expertise, should ease the implementation of the NWI and enhance the benefits achieved.

1 Social values and adaptive governance

The central insight of public choice theory is that collective choices expressed through the evolution of government policy reflect both the underlying values of constituencies and the institutional processes through which decisions are made. The emerging notion of adaptive governance highlights that these processes are shaped by experience and understanding, affecting attitudes and our understanding of the long-term consequences of specific patterns of resource management (Ostrom 2005). In its most general form, this literature argues that achieving improvements in institutional arrangements requires a sound understanding of environmental processes, the identification of feasible management responses (including new technologies), and the ability to craft policies or institutional arrangements that are perceived to be of value by key constituencies (Hatfield Dodds et al. 2006). While there are examples of institutional arrangements that have achieved sustainable water management over hundreds of years (Lansing & Kremer 1993), it is evident that growth in population, irrigation water demands, the expanding metropolitan footprint and climate change have placed strains on institutional structures trying to deal with increasingly complex problems where human needs can often collide. Institutions need to adapt to keep pace.

Contested values present both well-known challenges and less-recognised opportunities. The presumption that communities are polarised tends to support a 'crash-through' approach to reform, guided by distillation of *the* goal to be achieved. Lack of broad-based support makes this a fragile strategy – at least in democratic nations – given the depth of passion aroused by water issues. Recognition of multiple currencies of value allows a more nuanced approach, which seeks to build a coalition of support through identifying packages of action that give rise to outcomes of value to different constituencies. This approach focuses on the potential to expand the negotiation space and craft win–win solutions, rather than framing the entire process in terms of trade-offs between opposing values.

Values and passions about water run deep, and water resources cannot be separated from the society in which they are embedded. There is often a disconnection, however, between the goals of water management professionals and the communities that they serve. Management agencies and policy advisors tend to emphasise efficiency, control and industry outcomes. While these are important, the general community tends to view reform proposals primarily in terms of fairness and distributional impacts – revolving around the 'distribution of benefits and the costs of services, and who pays, and the distribution of risks ... who is vulnerable, and to what degree' (Delli Priscoli & Llamas 2001: 43). These underlying concerns provide fertile ground for disaffected groups and disadvantaged interests seeking to block worthwhile change. Overcoming this disconnection, and seeking to communicate the impacts of reform in the currencies of value to communities, could make a significant contribution to Australian water reform. However, the ability of our water agencies to deal with these issues is underdeveloped, with limited capacity to integrate expert and local knowledge and social science expertise.

2 The role of communication in Australian water resources planning and management

It is now passé to state that communication is a two-way process. This is widely acknowledged even within the water industry. It is achieving this communication or conducting a simple conversation with the wider community that is difficult to implement, despite the report of some successful public involvements. It is often considered that there are limitations to community input when it comes to decision-making, and for this reason government should be careful to moderate the influence of the community. Tacit reservations relate to an implicit belief that specialised qualifications or political judgment are required to make 'balanced' decisions in the public interest. More covertly, there is the concern that the involvement of the public may mean a diminution of the quality of the decision and the erosion of professional control. This section uses Australian data to examine both the input and the control issues.

There has been a series of studies of how the community perceives fairness in water allocation both in procedural and distributive terms. Depending on the allocation problem, there is systematic evidence that the community can apply many of the traditional 'academic' approaches of philosophers, economists and social scientists in lay language at both national and regional levels (Syme et al. 1999). These fairness criteria can be used to design decision-making processes in difficult allocation problems which clearly address the incorporation of overall societal values.

Similarly, there have been a number of demonstrations that the community can and is willing to make trade-off judgments for water decision-making. These have included setting service levels for urban water reliability issues (Hatton MacDonald et al. 2005), judging the desirability of investment in urban water aesthetics upgrades and expressing trade-offs for the allocation of water for environmental flows (Nancarrow & Syme 2001). The approaches have varied from simple focus groups, to structured workshops, to challenge survey techniques, to contingent valuation experiments and questionnaires and choice modelling surveys. The trade-offs made in these studies have been expressed in *utilitarian* terms (is the level of investment required by the utility worth it in terms of the product delivered?), in *procedural* terms (has the decision-making process been adequately implemented for an informed decision to be made?) or, as indicated above, in regard to *ethics* or *morality*. Given the opportunity to make judgments in water issues, the community can make reasoned preferences. Such analyses are not routinely conducted. This is a pity as the water agency's intuitive judgments about the community's expectations are often inflated.

In relation to planning criteria, there have been many studies using workshop or other procedures to provide criteria for assessing alternatives for decision-making. Planning budgets and other multi-criteria techniques are easy to run with the public in workshop settings. But are such exercises worthwhile? An excellent demonstration that they are was observed in the Wastewater 2040 futures exercise run by the Water Corporation (Water Authority of Western Australia 1995). A series of structured workshops were conducted in Perth suburbs of widely varying socio-economic status and in regional south-west communities. The workshops were successful in eliciting a series of evaluation criteria for assessing alternative futures. Engineers expressed some concern that the criteria might not be equivalent to those generated by professionals, and that a specialist workshop should be run. A more sophisticated workshop was conducted with planners and engineers. The evaluation criteria from this workshop, although expressed in more sophisticated language, were almost identical to those derived by the public (Syme et al. 1994). The quality of the public input was therefore sufficient to validate the process carried out by the Water Corporation during the remainder of the planning.

It seems, then, that the community can make judgments with ethics in mind, and can make useful trade-off judgments to enlighten planning, as well as provide meaningful evaluative criteria to judge between alternatives. The question then becomes: are we getting representative opinion on these issues, or is this information simply from a self-selected group of interested people? Many have had concerns about public involvement programs because it has been assumed that the higher socio-economic and special interest groups are likely to dominate and thus representative feedback will not be obtained. Consequently, input has to be 'second-guessed' when the public interest needs consideration. An alternative view is that representative input is not obtained because the public involvement processes are generally designed around procedures unintentionally created by middle-class professionals for like-minded and situated individuals.

A prospective study was designed to test the latter view in relation to water issues of particular concern to three Australian capital cities. People were asked to attend a workshop on the topic, and sent a personal invitation. They were assured that they had the capacity to contribute, that the process would be fun and that their opinions were valuable. About 15% of those who said they would like to come to the workshop attended. There was no difference in the likelihood of attendance in terms of age, income and education (Syme & Nancarrow 1992). The authors concluded that there was a potentially large demand for public involvement in water resources management. The Australian community does expect communication from the water industry but also wants to obtain feedback from it.

Having established a conversation, do people want to 'take charge'? The perception that they do may be why Australia appears to have a mismatch between the espoused government commitment to public involvement, and actual performance. But these fears appear to be groundless. This is demonstrated in Figure 2.1 (from McCreddin et al. 1997) which shows considerable agreement between professionals and the public about what should be achieved. The x-axis is Arnstein's ladder of influence (Arnstein 1969). The bar chart clearly shows that while the current level of public involvement was less than that desired by the community, the preferred level of power was partnership rather than community domination. It is interesting that the professionals also preferred such an arrangement.

Community members were also able to define what they meant as a partnership. The characteristics nominated for partnership bear a remarkable resemblance to the social psychology based theory of 'procedural justice' and the ethic espoused in the 'fairness' research outlined above. In essence, this involved the setting of the planning criteria and the planner

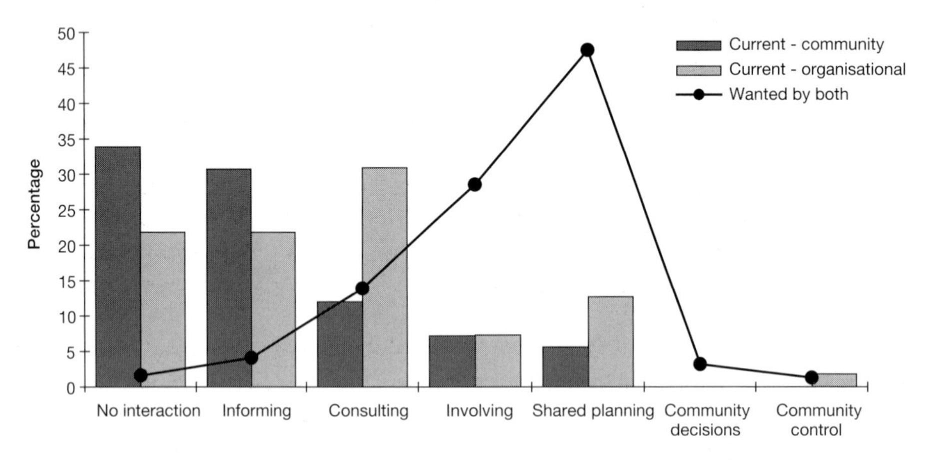

Figure 2.1 Perceptions of current community roles and preferred influence by the community and public servants in relation to water allocation. The single line shows what is wanted by both parties, because there was virtually no difference.
Source: McCreddin et al. (1997).

deciding on the best way to meet those criteria. The community is aware that they will not agree on all issues and that there will be conflict that requires careful management. Community members also acknowledged the value of expert knowledge and the benefits of an external view of local issues.

However, having established that there can be breadth and depth to 'partnership' conversations, the remaining issue is that of ensuring we are clear about what we are focusing on. Although this seems obvious, sometimes it may not be. Often the answer to a policy issue depends on the way the question is presented or framed. For example, it has been demonstrated that the compensation required to offset a loss is at least double that associated with an equivalent gain (Tversky & Kahneman 1991). The significance of this has been exemplified by Hatfield-Dodds and Jollands (2006), in a cross-Tasman study of willingness to accept climate change policy. Essentially there was significantly less support for a proposed policy if it was phrased in loss terms rather than forgone income. Since water reform incorporates public involvement, these findings create an ongoing challenge. It is important that the values implicit in alternative approaches are made clear in public involvement programs.

3 Values, attitudes and behaviour

There is a tendency in Australian debate to approach our communications with the assumption that people with different interests will have disparate values and attitudes. For this reason, this section examines what is known about differences and similarities in water-related attitudes among urban and rural communities in the context of value systems, water-using behaviour, water-conserving behaviour and responsiveness to innovation.

3.1 The urban community

Studies have regularly used multivariate statistical techniques to segment their consumers. For instance, there is a standard 'ways people think about water' inventory (Nancarrow et al. 1996–97), which can provide a method for observing changes in patterns of thinking over time and between cities. This covers contexts from the 'rights to water' and its environmental and aesthetic significance to aspects of water as a domestic commodity. For example, the development of this inventory with respondents from three Australian cities found that there were four basic clusters of people. This was replicated in a separate study.

1. **Self-interested** (33%) [20%] people with a general lack of strong thought about water in any context other than 'their rights'.
2. **Earthy** (29%) [23%] individuals who generally thought about water in its aesthetic, conservation and utility contexts but not in terms of additives or water rights contexts.
3. **Environmentalists** (18%) [32%] who were primarily concerned about water in terms of aesthetics and conservation rather than utility or water rights contexts.
4. **Service-oriented** (20%) [25%] people who generally thought about water in all its contexts, but more in terms of rights, additives and utility and least in the conservation context.

The approximate percentages in each cluster in the early 1990s (in round brackets) show that the largest segment was most concerned about rights in regard to water services, although the group was not hugely intererested in water issues. About half the community at this point was thinking about the wider roles of water in aesthetic and environmental terms. The percentages in square brackets were derived from the same questions from the second Perth domestic water use study (Loh & Coghlan 2003), held in 1999–2000, from which very similar clusters emerged. It can be seen that the community was thinking more of water conservation and was more interested in water than before.

The two Perth domestic water use studies conducted in the early 1980s and at the turn of the century (Water Authority of Western Australia 1983; Loh & Coghlan 2003) provide very detailed bases on which to predict water use. In the first study, total household consumption could be predicted from people's attitudes towards the recreational value of their garden, the perceived contribution of the garden to the resale value of the house, and concern about the price of water. Thus outdoor lifestyle, price and investment were prime motivators of use. At this stage attitudes towards conservation did not predict household consumption. The first study led to improved measures used in the second study. In that study, outdoor lifestyle was again a predictor of water use, as was attitude to price. A new variable also emerged – attachment to home as an environment. That is, 'home bodies' were greater users of water and attached more significance to the garden than those who spent more time away from home.

The second study also developed a model of external water use. This is important as it is more elastic than internal water use and is highly affected by restrictions. The model (Syme et al. 2004) showed that block size, income, garden condition and irrigation technology were all predictors of external water use. However, lifestyle, garden interest, outdoor recreation and conservation attitudes largely independently also explained external water use. Since about half of external water use was explained by attitudes and lifestyle, water managers need to carefully assess the social costs of restrictions. Fortunately the model provides a means of doing so.

A number of studies have assessed the determinants of water conservation in urban householders. There have been increasingly positive attitudes towards conservation over time. Kantola et al. (1982) showed both social norms and individual attitudes (such as attitudes towards price) affected intentions to conserve. For instance, Kantola et al. (1984) showed that the most effective form of message was a 'citizen's duty' one. Finally, Gregory and di Leo (2003) found that internal water use can be viewed as an habitual behaviour that is not susceptible to easy change. This is why the best means for encouraging indoor conservation are technology measures rather than persuasion (Syme et al. 2000).

Research so far indicates that pricing policies introduced to curb demand should be instituted in a climate of positive community attitudes if their potential is to be realised. Price elasticity among high-income consumers is low but may be enhanced by a positive social norm. An additional factor in the urban community relates to the acceptance of alternative modes of service delivery, and water reform will demand continued analysis of the ways in which water services are provided and where the water will come from. This will become urgent as recycling and more localised systems are considered. For this reason, Leviston et al. (2006) have developed a model to measure public assessment of a wide variety of alternative delivery systems (Figure 2.2).

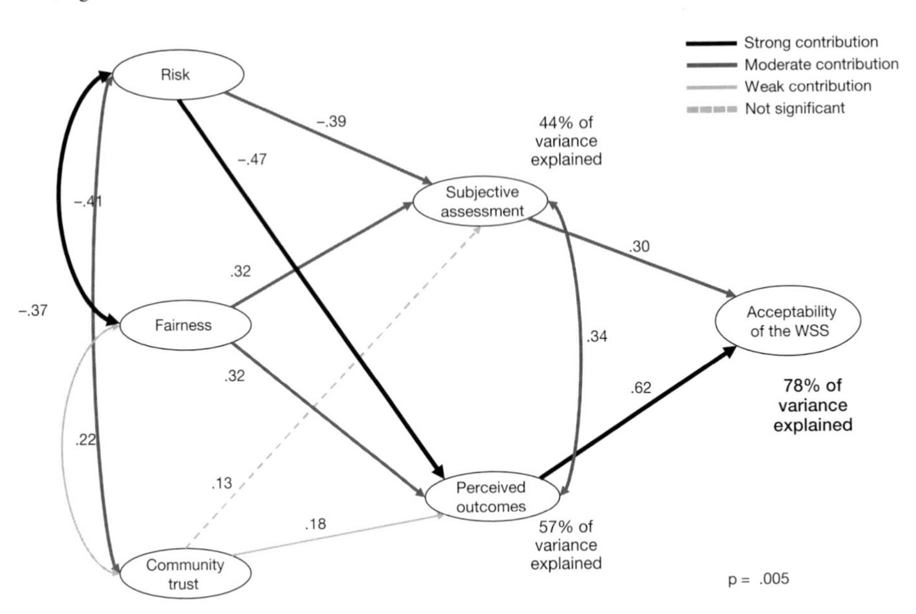

Figure 2.2 A model of acceptance of a new water supply system (an innovative stormwater-based system) in Brisbane. The components of the model, while varying in their influence, are constant for a wide variety of options throughout Australia.
Source: Leviston et al. (2006).

In addition to the personal benefits side of the equation, variables such as trust in institutions, equity and fairness and risk all significantly influence the judgment of acceptability. The wide range of service delivery options found to be acceptable by the urban community indicate that there will be considerable acceptance of innovation – if the basics of communication outlined above are met. Together with the finding discussed in Section 3 (see McCreddin et al. 1997: 37) that most community members desire input and consultation about water management objectives (rather than the detail of management and operational decision-making) this suggests that genuine consultation is often essential to produce outcomes considered acceptable and 'of value' to communities.

There is growing interest in water conservation but there are consistent differences between segments of the urban community in relation to water. All segments need to be considered in planning. It is important to consider quality of life issues associated with urban water use. Internal use is probably best changed through technology. There is considerable acceptance by the community of 'new' water delivery systems but there is a need to consider more than just household benefits. If innovation is to succeed, issues relating to trust, fairness and risk perception are also important determinants of acceptability of new systems. Fortunately we have a variety of social tools for analysing such issues and assisting in harnessing the public's ability to participate in decision-making for water reform.

3.2 Rural water users

In social terms the rural sector has been less studied specifically in terms of water use. Water use has often been seen as one of a suite of management behaviours, which have been examined by authors such as Mayberry and Crase (2001). There appears to be little literature on predicting water-conserving investment and behaviours, and even less literature on predicting acceptance of alternative modes of service delivery for the rural sector. Not surprisingly, there is a relative paucity of understanding and modelling of water use behaviours and alternative delivery systems for water.

Nevertheless, there has been a large literature within Australia on the development of farmer typologies in management based on location, crop type, expert judgment and personalities. This has been reviewed by Emtage et al. (2006). It is important that these insights are applied to water use in the greater detail now apparent for the urban sphere. Moreover, much of the farmer typology literature has arisen because there are large individual and situational factors that affect behaviour and there is a need to develop generalities for policy-makers. There has been an attempt to provide a systematic attitude and behavioural model for water use for irrigators in South Australia. The model (Leviston et al. 2005) did not explain a large amount of water use. Nevertheless, trust in the information provided by agencies and evaluation of using less water on production were important. In general, though, there would seem to be a need for more research in this area.

Summary

It would seem that specific behavioural and preference modelling has the potential to greatly facilitate water reform in both urban and rural contexts. It is important to note there is a need to understand individuals' water decision-making in relation to personal costs and benefits and in terms of their wider social values (or, in the rural setting, personalities). Behaviours can be largely influenced by institutional factors and issues such as trust, risk and fairness are likely to be important definers of progress in this area.

4 Public involvement in policy and planning frameworks

Water reform has emphasised the need for community involvement, which implies a need for detailed planning of the philosophy, roles, responsibilities and the process over time to support effective involvement. In rural areas there has been some monitoring of the social processes required for successful Landcare programs at a relatively localised level (e.g. Curtis 2003). This knowledge can help planning for future inclusion of a government–community partnership in the management of water. Nevertheless, water reform will require input into decision-making as well as action on the ground. There are strong arguments in support of public involvement in the development and articulation of ethical principles and policy parameters and it would be wise to delineate the scope and processes for public input, and the role of this input in water planning. There are also some challenges in terms of the support necessary for maintaining voluntary input into the future, given resource challenges and population age and distribution. In a recent futures study of the long-term role of volunteerism in catchment and natural resources management, Johnston et al. (2006) drew up twelve recommendations. Six were regarded as fundamentals that needed to be met if successful public involvement was to develop and be maintained in the long term; the report demonstrates that several of the vital ingredients for maintaining resilient volunteerism are missing.[1]

The broad thrust of the 1994 water reforms are generally interpreted in terms of achieving triple bottom line sustainability by using markets to move water to the use of highest economic value, while protecting the environment by limiting human consumptive use (see Smith 1998: 271). The 'social' bottom line was given little emphasis in the early period of reform, the main emphasis being on delivering changes within 'social constraints'. These constraints were not explicitly defined, although there was to be emphasis on consultation and public education. Importantly, culture as an input to water resource policy has been given little or no substantive attention. The social goals of water reform thus appear to have been restricted to avoiding unacceptably negative social impacts. It was not until recently that COAG began to examine social issues in a systematic fashion. With reform already gaining significant momentum in the environmental and economic domains, it is important to deepen our understanding of social opportunities and processes, particularly where there may be risks of irreversible or unnecessary adverse social impacts.

The recognition of the need for social planning in water resources management is not novel to Australia, nor are the problems in attaining it. To realise the extent of these problems we need only remember the Australian soldier-settlement schemes after the First World War, when many returned soldiers were settled in irrigation areas as new farmers. There has long been an understanding that water can be a direct determinant of many social benefits. The use of water for social aims has disappeared in preference to modern economic rationalism and the tendency for governments to refrain from intervening in society generally. Nevertheless, best practice in natural resource management demands that social considerations have a systematic and high profile, both in terms of outcomes and decision-making processes.

One of the issues raised by Australian water reform processes is that the emphasis on the expansion of water markets may have unintentionally framed the reform dialogue in terms of individual self-interest (Miller 1999) and encouraged a discourse which assesses outcomes primarily in terms of economic criteria rather than a balance between social, environmental and economic considerations. This culture has permeated many institutions, so it is not surprising that there is less attention to social or ethical principles that might govern water decision-making. This contrasts with nations such as South Africa and Indonesia, where social considerations and community roles and responsibilities are explicit (see Chapter 11). One of the key challenges in Australian water management is to create such principles and combine them with economic and environmental considerations to develop a holistic approach to water management. This will require a program of national and inclusive discussions to recreate the social purpose of water management. This will be a particular challenge in the attempts in both urban and rural environments to manage water in catchment and regional governance contexts as promoted by the National Heritage Trust and National Action Plan programs (http://www.nht.gov.au/). Incorporation of societal objectives in catchment and water plans will require close partnerships with the community (see Chapter 5). This will be particularly the case in rural areas where traditional local governance and volunteer community contributions are being forced to evolve.

The problems of governance in cities are becoming more complex. There is a growing realisation of the need to integrate the provision for environmental services with utilities and social services. In the past this has been inhibited by the largely separate disciplinary organisational bases of state and municipal governments. Engineers have provided physical infrastructure, social workers and social services in largely autonomous departments have been limited to sectoral expertise and thinking. The need to create capacity in holistic water planning in Australia is now urgent.

This evolution towards integrated planning, economic rationalism and community involvement creates opportunities for change but also creates challenges for local democracy and existing volunteer community infrastructure. An important question to consider in this context is where social and economic values are necessarily in conflict.

Our personal observation of water management and water reform is that professional managers and policy advisors often resist genuine community engagement and attention to social values and equity issues. This resistance is often related to a view that 'equity is in the eye of the beholder', thus there is no coherent set of social values or equity principles to be satisfied. We agree that there are multiple dimensions of equity and social values, but our own research also shows that a number of coherent generalisations about these topics hold true across most issues and population subgroups. We consider this coherence provides a basis for undertaking community engagement in ways that would help build support for worthwhile reform and enhance the outcomes achieved, more than an expert-led isolationist approach.

Our second observation is that the strong emphasis on fairness frequently relates to issues of procedural justice, such as the consideration given to impacts, the engagement of affected parties and their ability to voice concerns, and the degree to which arrangements seek to minimise adverse impacts. These issues are routine dimensions of good policy development and are generally regarded as adding value to reform. An additional nuance from social research is that respectful and honest engagement is widely considered to be intrinsically valuable, as well as instrumentally useful. Last, it should be noted that half-hearted community engagement is likely to generate additional mistrust and resistance to change, implying that there should be clear and open consultation about which aspects of a decision are open to discussion, and which aspects are effectively already decided.

Conclusion

While this summary has shown there is room for further social research, there is ample evidence that the community has the capacity to contribute significantly to water reform. More systematic community engagement, drawing on our national social science capacity and research insights, would strengthen and accelerate the implementation of the NWI and enhance the benefits achieved. Community contribution can also be enhanced through well-based education and engagement programs that motivate and inform water conservation. This would build on growing community interest in engagement with water management and conservation at the individual, business and catchment levels. Unfortunately, Australia has been unsystematic in incorporating community views and knowledge on issues relating to demand management, source development and water allocation issues. Many community and stakeholder attitudes are founded on perceptions of fairness and trust between water interests and in regard to institutions. Institutions need to be developed to promote adaptive learning and community–government partnerships. To do this, there must be firm underpinning of the volunteer contribution and social involvement with water planning, as suggested by the futures study. The prerequisites for future viability of community-driven structures are underdeveloped. This needs to be rectified if effective partnerships are to occur.

This is not rocket science. Rather than fiddling with ad hoc local approaches, delivering the full promise of water reform will require real commitment to building community support through respectful engagement and attention to water reform values and outcomes. This in turn will require development of a set of socially supported principles of water management outcomes, underpinned by community-sensitive water plans. In this way we can harness people's goodwill to develop truly sustainable water management.

References

Arnstein SR (1969). A ladder of citizen participation. *American Institute of Planners Journal* 35, 216–224.

Curtis A (2003). The landcare experience. In *Managing Australia's environment* (eds S Dovers & S Wild River), pp. 442–460. Federation Press: Canberra.

Delli Priscoli J (1998). Water and civilisation: using history to reframe water policy debates and build new ecological realism. *Water Policy* **1**, 623–636.

Delli Priscoli J & Llamas MR (2001). International perspective on ethical dimensions in the water industry. In *Navigating rough waters: ethical issues in the water industry* (eds CK Davis & RE McGinn), pp 41–64. American Water Works Association: Denver, CO.

Emtage N, Herbohn J & Harrison S (2006). Landholder typologies used in the development of natural resource management programs in Australia. *Australasian Journal of Environmental Management* **13**, 79–94.

Gregory GD & di Leo M (2003). Repeated behavior and environmental psychology: the role of personal involvement and habit formation in explaining water consumption. *Journal of Applied Social Psychology* **33**, 1261–1296.

Hatfield-Dodds S & Jollands N (2006). Issues in communicating the impacts of climate change policy options. Final Project Report to New Zealand Centre for Ecological Economics. June 2006. CSIRO: Canberra.

Hatfield-Dodds S, Nelson R & Cook D (2006). *Adaptive governance: an introduction, and implications for public policy.* CSIRO Working Paper: Canberra.

Hatton Macdonald D, Barnes J, Bennett J & Young M (2005). Willingness to pay for higher customer service standards in urban water supply. *Journal of the American Water Works Association* **41**, 719–728.

Johnston C, Green M, Stephens M, Syme GJ & Nancarrow BE (2006). *Volunteerism, democracy, administration and the evolution of future landscapes: looking ahead.* CSIRO Land and Water Client Report: Perth.

Kantola SJ, Syme GJ & Campbell NA (1982). The role of individual differences and external variables in a test of the sufficiency of Fishbein's model to explain behavioural intentions to conserve water. *Journal of Applied Social Psychology* **12** (10), 70–83.

Kantola SJ, Syme GJ & Nesdale AR (1984). An informational analysis approach incorporating Fishbein's model in a test of the effects of appraised severity and efficacy in promoting water conservation. *Journal of Applied Social Psychology* **13**, 164–182.

Lansing JS & Kremer JN (1993). Emergent properties of Balinese water networks: co-adaptation on a rugged fitness landscape. *American Anthropologist* **91**, 97–114.

Leviston Z, Porter NB & Nancarrow BE (2006). *Interpreting householder preferences to evaluate water supply systems.* Stage 3. Report to CSIRO Water for a Healthy Country. CSIRO Land and Water: Perth.

Leviston Z, Porter NB, Jorgensen BS, Nancarrow BE & Bates LE (2005). *Towards sustainable irrigation practices: understanding the irrigator.* A case study in the Riverland – South Australia. CSIRO Land and Water: Perth.

Loh M & Coghlan P (2003). Domestic water use study: In *Perth, Western Australia 1998–2001.* Water Corporation, Perth.

Mayberry D & Crase L (2001). *Social research to underpin the regional catchment plan implementation for the NSW Murray Catchment.* Report for the NSW DLWC.

McCreddin JA, Syme GJ, Nancarrow BE & George DR (1997) *Developing fair and equitable land and water allocation in near urban locations: principles, processes and decision making.* Report to the National Landcare Program. DWR Consultancy Report No. 96–60.

Miller D (1999). The norm of self interest. *American Psychologist* **54**, 1053–1060.

Nancarrow BE & Syme GJ (2001). *River Murray environmental flows and water quality project: stakeholder profiling study.* Report to the Murray-Darling Basin Commission. CSIRO Land and Water Consultancy Report. Published by the Murray-Darling Basin Commission, 2001, ISBN 1 876830 23 9.

Nancarrow BE, Smith LM & Syme GJ (1996–97). The ways people think about water. *Journal of Environmental Systems* **25** (1), 15–27.

National Heritage Trust (2007) <http://www.nht.gov.au/>

Ostrom E (2005). *Understanding institutional diversity.* Princeton University Press: Princeton.

Smith DI (1998). *Water in Australia.* Oxford University Press: Oxford.

Strang V (2004). *The meaning of water.* Berg Publishing: Oxford.

Syme GJ & Nancarrow BE (1992). Predicting public involvement in urban water management and planning. *Environment and Behavior* **24** (6), 738–758.

Syme GJ, Nancarrow BE & McCreddin JA (1999). Defining the components of fairness in the allocation of water to environmental and human uses. *Journal of Environmental Management* **57**, 51–70.

Syme GJ, Nancarrow BE & Seligman C (2000). The evaluation of information campaigns to promote voluntary household water conservation. *Evaluation Review* **24** (6), 539–578.

Syme GJ, Nancarrow BE, Casella F & Butterworth JE (1994). *Wastewater 2040 community involvement program: process evaluation.* Report to the Water Authority of WA. DWR Consultancy Report No. 94/44.

Syme GJ, Shao Q, Po M & Campbell E (2004). Predicting and understanding home garden water use. *Landscape and Urban Planning* **68**, 121–128.

Tversky A & Kahneman D (1991). Loss aversion in riskless choice: a reference-dependent model. *Quarterly Journal of Economics* **106** (4), 1039–1061. Reproduced in *Choices, values and frames* (eds D Kahneman & A Tversky, 2000). Cambridge University Press: Cambridge.

Water Authority of Western Australia (1983). *Domestic water use in Perth, Western Australia.* WAWA: Perth.

Water Authority of Western Australia (1995). *Wastewater 2040 strategy for the Perth region.* WAWA: Perth.

Endnote

1 (a) Catchment-based governance should be continued and integrated within existing government structures. Governance arrangements should ensure a holistic systems approach to environmental management so that relative powers match required accountabilities and responsibilities.

(b) The capacity of current legislative frameworks to address cumulative environmental impacts should be investigated and reformed to support catchment-based governance to meet environmental targets in accredited plans.
(c) One of the greatest vulnerabilities to ongoing volunteerism in catchment management is the stop-start nature of funding. To create certainty and sustainable volunteer-based structures, natural resource management must become a legitimate part of Australia's GDP and be funded by the federal government in the same way as health and education. In turn, catchment management bodies should not be dependent solely on one source of funding; they should look to securing long-term support from and partnerships with the corporate and industrial sectors.

(d) All levels of governance must recognise that investment in social capital is beneficial for the nation, but it does not automatically occur and nor is it 'free'. Cooperative behaviour and volunteering, as an inherent part of human society, must be deliberately fostered and supported in long-term policy at all levels of governance.
(e) While there are differing views on whether statutory backing is required for effective natural resource management, it is essential for the future of volunteerism that issues such as representativeness and inclusiveness in catchment decision-making be assured. The basic principles of procedural justice, including equal opportunity, empowerment and capacity to participate, must be integral to catchment management planning, implementation and evaluation.
(f) Knowledge from experts – scientists and consultants at state and federal levels – must be combined with local knowledge and expertise at the catchment level. Participation in scientific investigations and projects, as well as catchment planning and problem-solving, should be integral to natural resource management.

Chapter 3

Indigenous perspectives in water management, reforms and implementation

Sue Jackson and Joe Morrison

The National Water Initiative (NWI) represents a substantial change from previous national water policy in at least one respect: it explicitly recognises the special character of Indigenous interests in water. Parties to the NWI have agreed that water access entitlements and planning frameworks should recognise Indigenous needs 'in relation to access and management' (para. 25(ix)). Indigenous access is to be achieved through planning processes that:

- include Indigenous representation in water planning, wherever possible
- incorporate Indigenous social, spiritual and customary objectives and strategies for achieving these objectives, wherever they can be developed
- take account of the possible existence of native title rights to water in the catchment or aquifer area
- potentially allocate water to native title holders
- account for any water allocated to native title holders for 'traditional cultural purposes' (paras 52–54).

Three other clauses relate specifically to Indigenous interests. First, water plans are to provide a statutory basis for achieving 'environmental and other public benefit outcomes', and these include 'Indigenous and cultural values' (para. 25 and Schedule B(ii)). Second, Schedule E to the NWI provides guidance for the preparation of water plans, including a requirement to consider Indigenous water use. Third, protection of certain Indigenous heritage values is included as a principle to guide the establishment of water trading rules. According to these principles, restrictions on extraction, diversion or use of water resulting from a trade can be made to manage 'features of major Indigenous, cultural heritage or spiritual significance' (Schedule E, para. 3(v)). The above provisions provide a much-needed impetus to addressing Indigenous people's customary requirements for water, as well as enhancing their participation in water management systems. Indigenous people have a large stake in water management arising from their long traditions of water management, customary land and resource rights, some of which are recognised by the common law, and an extensive and growing land base.[1] River valleys have been the main focus of Indigenous life for tens of thousands of years and water maintains a significant symbolic part in Indigenous social life, including contemporary identity. Widespread Indigenous disadvantage provides further cause for consideration of the socio-economic impacts of water reform on regional economies.

Before reviewing the existing knowledge base of value in meeting the challenges posed by the NWI, we wish to highlight three issues that may affect the degree to which Indigenous people benefit from the NWI, as well as the benefit to the broader nation derived from their participation in the water sector. First, and as acknowledged by McFarlane (2004), most of the NWI provisions are expressed in discretionary terms. Although this provides flexibility to suit a wide array of circumstances, impediments and competing priorities may hamper the extent to which Indigenous objectives are addressed. Close monitoring of water management planning processes and the impact of trading systems will be required to determine the extent to which this and other related requirements are satisfied. Second, water resource managers are given little guidance to address objectives relating to Indigenous access and involvement (see also Connell et al. 2005). Researchers, Indigenous groups and policy-makers will need to collaborate to overcome several key challenges that may impede progress. These include limited knowledge of how to address Indigenous water requirements and the degree of technical difficulty, lack of Indigenous and government agency capacity, and the impediments posed by uncertainty, contestation and lags in native title claims processes (e.g. NSW Healthy Rivers Commission 1999).

Third, implementation emphasis is squarely placed on protecting Indigenous customary values (which are construed as non-market values) and meeting legal requirements to protect native title. The NWI actions reflect statutory frameworks for native title and interpretations of Indigenous resource interests that are, by numerous academic accounts, 'insufficiently inclusive in their definition of water property' (Altman 2004a: 33). The property rights conferred by the *Native Title Act 1993* are only partial, covering customary use rights (Altman 2004a). Furthermore, the wording in the NWI, that water allocated to native holders for *traditional cultural purposes* (para. 54; emphasis added) is to be accounted for, suggests an intention to preclude commercial uses under the definition of native title rights. However, the absence of definition leaves some doubt as to what is intended. In light of Australian government commitments to overcome Indigenous disadvantage, it is significant that no explicit obligation is placed on the parties to utilise this market-based policy framework to advance Indigenous people's economic standing.

The difficulty in finding Indigenous commentary that was explicitly relevant to our task underlines two interrelated points that serve to qualify this chapter and, more importantly, reinforce the need for extensive Indigenous engagement in the roll-out of the NWI. We suggest that the absence of Indigenous commentary in published and grey literature indicates negligible awareness of the water reform agenda among the Indigenous population. This, in part, explains the very limited input from Indigenous people and organisations into water policy development noted in the literature (Altman 2004a; Jackson & O'Leary 2006; McKay 2002). Indigenous organisations did not participate in negotiating the terms of the NWI, nor does it appear that they have subsequently considered its implications or impacts. Seen from this perspective, elements of the NWI relating to community partnerships, knowledge and capacity-building assume great importance in addressing the social justice requirement for Indigenous participation.

1 Current knowledge and capacities, and adaptable knowledge

Palmer (1991) warns that generalisations considered to apply to all Australian Indigenous people are sometimes misleading. Diversity exists across river basins in land use, population and social priorities, as well as in forms of social organisation and resource management institutions. Similarities can be distilled, however, with Indigenous people sharing 'a desire to retain their identity, a belief in their right to their land, a desire to control their own affairs, and a desire to remove economic and social disadvantages', although the strategies employed to achieve those aims vary (Horton 1994: xx). Since the late 1990s the significance of water and rivers to Indigenous societies has grown as a topic of interest to researchers, lawyers engaged in native title processes, natural resource management groups, and water resource managers. Some attention has been given to the definition, explication and incorporation of Indigenous values relating to water in environmental planning and to Indigenous aspirations for resource governance in catchment-scale institutions (Jackson 2005, 2006; Morgan et al. in press; Weir 2006a). Yet Indigenous interests were not formally considered in water reform policy documents prepared during the 1990s (Tan 1997; Lingiari Foundation 2002), nor addressed in water resource law until 2000. Thus the field surveyed here is very recent and contains substantial knowledge gaps. It can be further characterised as geographically biased towards cultural analyses of northern Australian circumstances where there are fewer resource pressures, with much of the legal analysis focused on speculative discussion of the nature and extent of Indigenous rights. There is an observable disjuncture between the significant attention given to Indigenous economic rights in the academic literature and the content of resource management discourse which rarely addresses property rights issues, or economic opportunities arising from Indigenous access to water.

In the following section we review the existing and adaptable knowledge relevant to four implementation tasks distilled from the NWI. We chose them in order to avoid duplication of points common, or closely related, to a discussion of both native title rights and other Indigenous water use and management objectives:

- incorporation of Indigenous social, spiritual and customary (including heritage) objectives in water planning
- recognition of the possible existence of native title rights to water
- allocation of water to native title holders for traditional cultural purposes
- Indigenous representation in water planning.

1.1 Incorporation of Indigenous social, spiritual and customary objectives in water planning

An important prerequisite to meeting Indigenous water needs is a greater general awareness of Indigenous concepts of country, the nature and extent of Indigenous interests in water, and their relationship to other Indigenous values such as identity. Disciplines within the humanities and social sciences, such as anthropology, geography and environmental history, have sought to understand the ways in which various social groups conceptualise and engage with the non-human world, including the part played by culture, religion and scientific rationality in Western perspectives on nature. Since the early 1990s accounts of Indigenous environmental philosophy, cosmology, land tenure systems and management practice have contributed considerably to the academic and policy knowledge base, with the intention of improving collaborations and engagement between different knowledge and value systems (see Woenne-Green 1994; Rose 1995; Rose 1996, 1999; Langton 1998; Howitt 2001; Sharp 2002). Notwithstanding these contributions, the lack of awareness of Indigenous interests within natural resource management agencies still remains a serious capacity-building challenge (Larsen & Pannell 2005: 17; McFarlane 2004). This review will not address that literature, but it is important to reiterate that an exclusive focus on water use and management is antithetical to the holistic quality of Indigenous environmental perceptions and discourse that consistently emphasise connectivity and relationships between features and components of the socio-ecological system (Toussaint et al. 2005).

There is a growing body of literature explicitly documenting and analysing the ways in which Indigenous societies attribute meaning to water and the place of water in their formalised systems of knowledge and social institutions (Langton 2002, 2006; Strang 2001; Toussaint et al. 2005; Barber & Rumley 2003; Rose 2004; Jackson 2004, 2005). Much of this literature is drawn from ethnographic studies carried out in northern Australia, in regions such as the Kimberley, reflecting a geographical bias in our general knowledge of Indigenous societies (Maddock 1982). Water is examined as a feature of the Indigenous cultural landscape with significant attention devoted to the symbolic dimension of individual and group attachment to water. Northern studies describe and interpret stories relating to water represented in myth, painting, film and dance, and the local customary practices, beliefs and ideas associated with water (Toussaint et al. 2005). As well as examining this 'intellectual use' of water, as Trigger (1985) describes its symbolic, metaphorical and conceptual significance, the studies also reveal the material use of water according to Aboriginal custom. Many commentators refer to water's economic significance as a vital element underpinning the Indigenous harvest and intra-community distribution of aquatic life (e.g. Behrendt & Thompson 2004; Altman 2004a; Strang 2001). Historically, Indigenous interventions improved rates of harvest of certain species; for example, river flows were manipulated with the construction of fish traps, weirs and small dams in numerous Australian river systems (Tan 1997).

From this literature it is apparent that water plays a central role in Aboriginal societies. All studies reviewed here observe that groups conceptualise water sources and rivers as having derived from the actions of mythic beings during the Dreaming, when the world attained its present shape and the socio-cultural institutions governing water use were formed (Barber & Rumley 2003; Langton 2002; Toussaint et al. 2005; Yu 2000). Water's vitality is underscored in all accounts, including those from the Murray-Darling Basin (Morgan et al. in press). In contexts where resources or places are under pressure, there may be a tendency to focus on key places or sacred sites as people strive to retain their traditions (Kolig 1996). However, affiliations to water are much broader than those encompassed by the conventional cultural heritage paradigm: these humanitarian values relate to notions of sociality, sacredness, identity and life-giving (Jackson 2005, 2006). Within the literature, there are references to congruence between some Indigenous landscape constructions and valuations and those maintained by non-Indigenous groups (Strang 2001; Langton 2002; Goodall 2002; Windle & Rolf 2002). This highlights the need to understand better what processes are most influential in the formation and development of cultural meanings and environmental perspectives, and may provide an opportunity to build cross-cultural consensus around shared concepts, such as human–water interdependence. It may also lead to more collaborative approaches to water management (see Delli Priscolli 1998; Chapter 2, this volume).

A related challenge lies in identifying and understanding the means available to incorporate Indigenous social, spiritual and customary objectives in water planning. It is beyond the scope of this chapter to evaluate how well current planning processes are addressing Indigenous objectives. The very limited literature available describes the types of mechanisms and approaches developed in only a few jurisdictions. Some degree of critique, particularly with respect to the NSW water legislation, is provided by Behrendt and Thompson (2004; see also Douglas 2004; Tan 1997), and McFarlane (2004). Jackson's case study of the Northern Territory's Daly River (2005, 2006) reveals certain conceptual and technical challenges to addressing Indigenous interests in catchment and water planning. Indeed, we have no current overview of the various methods of identifying and incorporating Indigenous objectives within water planning. An ability to compare and evaluate the range of techniques would be ideal. However, Indigenous water management initiatives appear to show that matters of assessment technique are presently less important than the establishment of appropriate terms of engagement for partnerships. It is clear that most attention is being given to resource governance with Indigenous groups seeking to assert their rights, create inclusive processes and collaborative relationships based on recognition of cultural difference, including Indigenous law and custom. Once established, these 'co-management regimes' (Carlsson & Berkes 2005), as they have come to be known in the international literature on power-sharing in environmental management, become platforms for collaborative decision-making and problem-solving.

A number of concepts have been adapted or designed to recognise Indigenous values, most notably the 'cultural flow' concept emerging from contributions from Indigenous nations to the Living Murray Initiative (Morgan et al. 2004; Weir 2006a). There is also the 'cultural value' or 'beneficial use' concept, under the National Water Quality Management Strategy (see Jackson 2005). Indigenous responses to environmental flows policy, water access rights and trading of entitlements in the Murray-Darling Basin have been observed by Weir (2006a, 2006b) and Morgan et al. (in press). Indigenous-led initiatives include agreements between various nations and the Murray-Darling Basin Commission (or state governments and local government), cultural heritage management plans providing employment for Indigenous people, and co-management arrangements with conservation agencies (Morgan et al. 2004).

A comprehensive examination of the NSW approach provides one of the few studies of direct relevance to implementing the NWI (Behrendt & Thompson 2004). The translation of Indigenous spiritual values into legal rights and interests in NSW is described as being in its formative stages, and there remains a considerable amount of uncertainty as to the complete range of rights and interests to be recognised (Behrendt & Thompson 2004: 78). Behrendt and Thompson express doubt about the beneficial effects of the legislative framework, which is arguably the country's most comprehensive, and are critical of the low priority afforded Aboriginal interests compared to a range of competing interests, the unenforceable nature of policy initiatives and the minimal protection of native title rights and interests.

The extent to which NSW's native title provisions will satisfy native title requirements can only be determined on a case-by-case assessment, although it is significant that a desktop review of the thirty-five NSW water-sharing plans in operation reveals that only two have provided an allocation for native title.[2] Behrendt and Thompson (2004) repeat the point observed by other legal commentators (e.g. McKay 2002) that little satisfaction could be gained from an entitlement to extract water in the exercise of native title rights if there is insufficient water to extract or if it is not of a consumable quality. Interaction between Indigenous rights and needs and competing interests has surfaced in case law involving disputes over water and fisheries management in Canada, New Zealand and the US, and further attention will be given to the topic in the section below on impairment of native title.

The NSW *Water Management Act* also provides for the grant of specific-purpose licences to Aboriginal people or communities for cultural or commercial purposes in macro sharing plans. Conditions under which such licences can be granted are numerous and may be limiting Indigenous access. There is also provision for the granting of Aboriginal cultural licences for fifteen years under the NSW *Water Act 1912*.[3] An officer of the NSW Department of Natural Resources consulted during the course of this review was not aware of any licences being granted for either purpose (Miller, pers. comm.). One cultural access licence has recently been granted to an Indigenous group in the Murrumbidgee region (NSW Dept of Natural Resources 2006).

As the NWI allows trade restrictions to manage features of major Indigenous heritage significance, the field of heritage management also requires consideration. Similarly, as water bodies are generally considered of special religious significance by Indigenous people, heritage protection measures have frequently been sought for activities (see Table 3.1). National guidelines for the recognition of Indigenous heritage based upon local and international protocols (Australian Heritage Commission 2002) will be useful in this context, as will the extensive critique of heritage management policy and practice. Much has been written on methods for assessing significance, elucidating contemporary attachment to land, difficulties associated with revealing secret/sacred knowledge (Hancock 1995), the preoccupation with tangible heritage (Byrne et al. 2003; Weiner et al. 2002) and the adequacy of various statutes (Goldflam 1997; Behrendt & Thompson 2004).

Table 3.1 Water-based activities with potential to affect Indigenous practices and values (with key references)

Activities and impacts	Key references
Construction of dams, weirs and altered water levels, or draining water bodies. Effects felt on places of significance such as sacred sites, burial grounds, canoe trees, middens and fish traps.	O'Connor et al. (1989); Barber & Rumley (2003); Langton (2002); Behrendt & Thompson (2004); Windle & Rolfe (2002); Morgan et al. (in press); Yu (2000)
Recreational activities such as fishing, boating and water-skiing (e.g. Boobera Lagoon on the Barwon River in NSW). Disturbance to sites, competing resource use.	Behrendt & Thompson (2004); Jackson & O'Leary (2006)
Commercial fishing, shipping and port operations.	Smyth (2002); NAILSMA (2006)

We are not aware of any instances where heritage legislation has been applied to protect significant sites from the adverse effect of water trading. A Queensland study confirms that Aboriginal heritage issues have not been directly considered by water policy (Windle & Rolfe 2002). The indirect and cumulative effects of water use on Indigenous interests deserve further, albeit brief, comment. Assessing impacts on heritage values is likely to be difficult, as is the case with maritime activities (Campbell 2002), largely because of the cumulative dimension and the difficulty in proving causation where trans-boundary water impacts are concerned.[4] These characteristics of water use warrant further consideration of broad landscape approaches to impact assessment (English 2002). In NSW, for example, heritage values are being incorporated into environmental planning processes through cultural mapping methods, social impact assessment and planning instruments such as regional vegetation plans (English 2002). This trend is likely to be of considerable relevance to water planning.

1.2 Recognition of the possible existence of native title rights to water

Following a brief overview of the legal and policy analysis of Indigenous rights to access water under the umbrella of native title, we will discuss two issues prevalent in the literature: first, the partial nature of the property rights in water conferred to native title parties (which the High Court has characterised as unique, and presently accepted as non-exclusive and therefore not commercial); and second, the interactions between Indigenous interests and others within contexts that require resource-sharing.

Indigenous peoples' interests in land and water are recognised by two branches of law applicable in Australia. Prior to *Mabo* (No. 2),[5] some Indigenous groups had rights in relation to specific parcels of land and water because the Crown had granted those rights and titles under land rights legislation based on a model originating in the Northern Territory (Neate 2002). This legislation generally makes no mention of ownership or rights to inland waters, or indeed any other resource rights (O'Donnell 2004: 102; Altman 2004b). Notwithstanding, the material prepared for numerous land claims in the Northern Territory and other jurisdictions with statutory land claims processes, represents a rich source of reportage on Indigenous relationships to water (Toussaint et al. 2005; Tan 1997). In the *Mabo* decision, the High Court recognised for the first time the entitlements of Indigenous people to their traditional lands under their traditional laws (Neate 2002). According to O'Donnell (2004) there is now no doubt in general terms that Indigenous legal rights to inland waters exist in Australia. When the *Native Title Act 1993* was passed, the scope of native title was defined to include rights over waters[6] located within traditional estate boundaries. It confirms governmental ownership of water and minerals, while guaranteeing rights to customary use of resources for sustenance (hunting, gathering and fishing). A point emphasised by numerous analysts is that these customary uses are often dependent on inland water (Altman 2004a).[7] Most, if not all, legal commentators agree that uncertainty about the extent and nature of native title rights remains high, and there is sharp criticism of the lack of outcomes from native title processes (Strelein 2004).

Few published studies specifically describe how Indigenous people conceive of their rights to water and the customary rules governing land and water utilisation (see Langton 2002, 2006; Toussaint et al. 2004), although anthropological literature prepared for claims represents a largely untapped source of information. Some water resources were 'clearly considered capable of ownership' (Tan 1997), according to Aboriginal custom, and whether ownership of flowing water can be established will depend on the facts of a particular group's use of water. The primary question that remains unresolved is whether native title includes the right to 'exclusive possession, occupation, use and enjoyment of waters' (O'Donnell 2004). New Zealand and Canadian judicial decisions suggest only a limited degree of development, to allow for more contemporary forms of exploitation (Bartlett 1997). McFarlane (2004), like many others (O'Donnell 2004; Tan 1997; Altman 2004a), expects that native title rights to water will be non-exclusive and non-commercial in nature, allowing native title holders to use and take for domestic and subsistence purposes only. A right to protect sites or areas of significance that include waters has been recognised as a native title right (O'Donnell 2004). Such native title rights will be subject to existing laws and grants and are of a more limited nature than rights attached to land (Collings 2002).

Indigenous water policy literature gives considerable attention to economic rights, described by Campbell (2002: 174) as the right-holders' ability to enjoy benefits from property (Kaufmann 2004; Altman 2002, 2004a, 2004b; Langton 2002). There is a similar interest in better defining a right of Indigenous participation in commercial fisheries (e.g. Tsamenyi & Mfodwo 2000; Campbell 2002; Smyth 2002), carbon trading and a general Indigenous development right (Gibbs 2005; Altman 2004b). Other authors have compared Australia's situation to common law countries and found greater recognition of the relationship between land and livelihood in Canadian and US Indigenous people's claims (Behrendt & Thompson 2004; Langton & Palmer 2003; Getches 2005). Precedence created by the New Zealand Maori fisheries settlement is often cited.

Importantly, the NWI requires that water planning processes 'take account of the possible existence of native title rights to water' (para. 53), but does not articulate how this might be done. Presumably, it would encompass consideration of the effect of water use activities that may impair native title rights, as well as the potential to allocate water to title holders to ensure their continued enjoyment. In relation to impairment, one of the principal objects of the *Native Title Act 1993* is to recognise and protect native title. It provides a mechanism for permitting actions that may affect native title, such as granting water licences. Behrendt and Thompson (2004) discuss how these 'future acts' may affect Indigenous interests in rivers and water (e.g. whether they are compensable), noting that certain procedural rights of notification apply. Again, there is uncertainty as to how broadly the courts might construe the manner in which native title rights and interests might be affected. Behrendt and Thompson (2004) observe from overseas experience that the courts have taken a broader view 'than what some governments may prefer'. They detail the US example, where Indigenous people have enforced a level of access to water that ensures not only that their lifestyle be sustained but that they have sufficient water to pursue agricultural livelihoods. They and other authors argue that, despite different legal and historical circumstances, the principles upon which native title can be considered to be affected are arguably equally applicable to Australia's situation. One such principle is the implied right to water because it is an integral component in Native American interests in fishing (Behrendt & Thompson 2004; Lombardi 2004).

We have little guidance in determining how these issues might be resolved in Australia for the following reasons:

- the effects of expanding existing water markets and the creation of new water markets on Indigenous people are not known
- native title decisions to date have not considered the effect of unsustainable water use on Indigenous rights and interests
- we lack clarity on the consequences of failing to provide the procedural rights set out in the *Native Title Act*, or how compensation may be determined.

The uncertainty and potential for high transaction costs arising from conflict, compensation and disaffection between parties suggests that there should be inclusive processes that bring parties together to negotiate. Agreement-making is perhaps the most commonly referred to means of satisfying the range of co-existing interests in Australian water management (Altman 2004a; Sheehan 2001). Smith (1998), Edmunds (1998) and Langton and Palmer (2003; Langton et al. 2004) discuss the opportunities and challenges provided by a mechanism of the *Native Title Act*, the Indigenous Land Use Agreement, which can apply to water. The cited benefits include improvements to relationships between Indigenous people and other parties (Smith 1998), the establishment of co-management arrangements that meet environmental requirements, and economic and legal objectives (Goodman 2000; O'Faircheallaigh & Corbett 2005).

1.3 Allocation of water to native title holders

The NWI envisages the situation in which water may need to be allocated to meet certain Indigenous requirements: Indigenous use, landscape features of value and native title. There are substantial conceptual and technical difficulties facing water resource managers seeking to calculate and allocate water to meet these needs. We can refer here only briefly to this outstanding knowledge gap. We are aware of few attempts to identify an Indigenous share in an allocation process, although some preliminary consideration has been given to defining the problem from an Indigenous perspective and critically examining the culture of water administration (Jackson 2005, 2006; Weir 2006a). This work reveals the need for greater clarity in conceptualising the nature of Indigenous water uses or needs, as well as techniques for elucidating the relationship between flow and values. In NSW and the Northern Territory, at least until recently, the approach was to assume Indigenous requirements would be served through a surrogate environmental flow (Jackson & Langton 2006; Jackson 2005; Cohen 2004). Indigenous groups participating in environmental water allocation processes consistently express a strong desire to ensure adequate environmental flows (see Anon. 2003; Miller, pers. comm.) but questions remain: Are environmental flows sufficient? If not, is it extra or 'different' water that is required and how is it to be determined and accounted for?

From descriptions of the NSW water-sharing plans it is not clear how the Indigenous specific allocations were set and what process will be followed when evaluating these arbitrary allocations to determine if they are meeting Indigenous needs. The Apsley water-sharing plan, for example, is one of two plans allocating water to meet native title requirements, in this instance 0.01 ML/day to an Aboriginal community residing by a river. This amount was determined using a formula based on per capita residential water use rather than considerations relating to spiritual, economic or cultural objectives. The community had a water frontage and was therefore entitled to a basic landholder right to water (domestic and stock purposes) (McKay 2002), revealing the lack of distinction between basic and native title water rights (Miller, pers. comm.).

It is worth considering the value of determining a separate 'cultural flow', as proposed by the Murray-Darling nations (Anon. 2003). According to this proposal, a water allocation should be available 'to each Indigenous nation to enable them to exercise their custodial responsibilities to care for the river system' (Anon. 2003). Autonomy among the individual nations would entitle each to decide its use: whether the allocation should contribute to environmental flows or help generate greater economic independence (Anon. 2003: 7). In such a case, water might be put to different uses under the auspices of a 'cultural flow' (Weir 2006b), although it is unclear how this change in use would be accounted for. Based upon our experience in northern Australia, where many rivers remain unregulated and in relatively good health, we question the value of the concept.

First, it is plausible that a water regime meeting environmental requirements will sustain certain Indigenous values (Jackson 2005), assuming that government agencies support Indigenous management activity. Second, it is difficult to conceive of a practicable and coherent way to specify the quantity of water required to meet the needs of 'culture' to the extent implicit in the cultural flow idea (Jackson 2006). The Indigenous uses involving water are hunting, fishing, gathering, cultural and spiritual activities. In current terminology these tend to be considered as non-extractive, because people do not need to 'take' water to enjoy the activity. Of course hunting and fishing are dependent on healthy ecosystems, but if they are diminished they are unlikely to be significantly improved by 'watering' via a withdrawal separate to an environmental allocation. In the Murrumbidgee case, where a cultural access licence has been granted, it is acknowledged that the cultural allocation 'works like an irrigation licence' in watering a wetland owned by the Aboriginal community. It does not recreate the fish breeding cycle affected by river regulation (Christa Hay, pers. comm.). Finally, there may be an implied right to water and environmental protection sufficient to sustain customary activities, such as fishing, but whether there will be significant Indigenous interest in abstractly separating these uses and quantifying the water required to satisfy them remains to be determined. Indigenous groups aspire to maintain the indivisible quality of the environmental relationships that sustain the underlying native title, and therefore a narrow and fragmented approach may be considered too academic.

1.4 Indigenous representation in water planning

Indigenous systems of customary law dictate that traditional land-owners have a substantive role in land and water management and resource regulation, and hence a particular interest in environmental governance structures (Morgan et al. in press; O'Faircheallaigh & Corbett 2005). Recent studies observe that Indigenous expectations of the extent to which they can directly participate in water management institutions are not being met (McFarlane 2004; Behrendt & Thompson 2004). Indigenous participation in a range of environmental management sectors has been described and barriers analysed (Baker et al. 2001; Lane 1997; Howitt 2001; Lane & Corbett 2005), and guidelines for the natural resource management sector have been produced (Smyth et al. 2004).[10] Attempts have been made to assess the level and nature of participation (see McGurk et al. 2006), although independent assessments of participatory natural resource management programs are not widespread. Notably few empirical studies have evaluated the effect of growing numbers of Indigenous reference groups supporting regional NRM bodies, or the quality of Indigenous participation in multi-stakeholder catchment groups (see Lane & Corbett 2005). Table 3.2 summarises research issues discussed in the literature on Indigenous participation in environmental management. These issues are likely to be pertinent to NWI implementation.

Table 3.2 Key references in literature on Indigenous participation in environmental management by issue of interest to social researchers

Issue or topic	Key references
Power imbalances and minority part played by Indigenous members of multi-stakeholder committees.	Behrendt & Thompson (2004); Whelan & Oliver (2005); Lane (1997); Lane & Corbett (2005); Howitt (1993)
Factors inhibiting acceptance of Indigenous modes of decision-making, representation, norms and mores, communication and accountability.	Lane (1997); Ross (1992); Howitt (2001); Paton et al. (2004)
The need for nuanced understanding of 'community' and distinguishing between the roles of traditional owners and other members of the Indigenous community.	Smyth et al. (2004); Baker et al. (2001); Morgan et al. (in press)
Lack of evaluation of catchment management authorities' performance in respect of obligations to Indigenous people.	Behrendt & Thompson (2004); Lane & Corbett (2005)
The existence of Indigenous protocols for representation and participation, as well as guidelines for engagement of Indigenous people in planning.	Morgan (in press); Behrendt & Thompson (2004); Baker et al. (2001); Ross (1992); Larsen & Pannell (2005); Jackson & O'Leary (2006)
The match between the scale of Indigenous social organisation, Indigenous notions of regionalism and catchment-scale units of management.	Jackson et al. (2005); Lane (1997); Lane & Corbett (2005)
Incompatibility between Indigenous knowledge and aspirations for holistic management and scientific knowledge and technical forms of rationality.	McFarlane (2004); O'Faircheallaigh & Corbett (2005); Langton (1998); Jackson (1997)
Need for empowerment of Indigenous people and capacity-building among non-Indigenous resource managers.	Howitt (1993); Lane (1997); Storrs et al. (2001)

2 Research gaps and opportunities

Given the potential for intense competition over common resources, literature from fisheries, forestry and water resource management contexts offer various models and approaches to building management systems that accommodate coexisting rights and interests and fulfil criteria for sustainable use. Appreciable international scholarly attention is being given to conflicts over water arising from the articulation of market mechanisms and local customary formulations of rights and equity (Bruns & Meinzen-Dick 2000; Molle 2004; Roth et al. 2005; Singh 2006). These water rights studies critically examine the contested nature of uneven access to water, its role as a commodity and crucial resource in rural livelihoods and its 'fundamental source of cultural meaning, identity and social identification' (Roth et al. 2005: 3). Much attention is given to irrigation systems in Asia and South America (Singh 2006), and therefore may be less directly relevant to the Australian context where there is minimal evidence that customary uses entailed control and regulation of water for agriculture (e.g. Victorian eel farms). Insights into notions such as legal pluralism, as developed in overseas water resource situations, may improve our analytical understanding of natural resource use, especially the normative dimensions of water control (Roth et al. 2005) and the highly complex recursive human relationships with water in which material uses, experiences and symbolic meanings are intertwined (Strang 2004).

In Australia, as native title cases are heard, more evidence of the customs and laws relating to water will be revealed. We may achieve greater clarity in the definition and articulation of Indigenous rights and interests compared with introduced property regimes. Meanwhile, there is an important role for social scientists and analysts of law and Indigenous policy, as well as Indigenous community organisations, to play in meeting water policy's fundamental information needs:

- document and analyse Indigenous institutions governing water management (including rules, norms, rights and responsibilities relating to water)
- assess the impact of the water reforms, particularly water trading, on Indigenous water use, economic patterns and wider social values, and identify points of tension or conflict between the customary and market sectors
- ascertain Indigenous aspirations for water management and use, including preferred development models, and ways of achieving greater participation. The interests of Indigenous groups living in urban centres should be included, as they have received even less attention than groups living in rural or remote areas
- investigate models and methods to incorporate Indigenous water resource management systems into emerging and evolving institutions, such as negotiated agreements.

Our understanding of the barriers and incentives to Indigenous participation in the water sector is inadequate. There is no overall system in place to monitor Indigenous involvement in water-based enterprises or water management institutions, making it impossible to accurately describe the present situation and access trends. A significant contribution to establishing a baseline for monitoring and evaluation would be made by reviewing each jurisdiction's legislative and policy framework to determine:

- the extent and quality of Indigenous engagement and participation in preparing plans, particularly their role in advisory or reference groups
- the extent to which the states are meeting other implementation tasks, such as allocating water to Indigenous uses, upon what basis and by what methods
- the extent to which states are identifying and assessing the effect of water use decisions on native title
- monitoring and evaluation systems
- the relative merits of the various participatory, deliberative and modelling techniques.

Cultural analyses have illuminated the way in which the contemporary consumption and distribution of resources is underpinned by beliefs, values and social relations, but we do not know of any studies that have analysed in detail or quantified the value of water resources to the contemporary Indigenous economy. Nor is enough known about Indigenous attitudes towards water conservation, efficiency or methods of cost recovery (Willis et al. 2004). Although the NWI does not specify Indigenous economic objectives among the interests requiring attention, Indigenous representatives and analysts of Indigenous policy frequently urge consideration of this aspect, at least in so far as competing water uses may impair the enjoyment of subsistence activities (Altman 2004a). This would require greater knowledge of water's value to the customary economy, the threats to continued realisation of this benefit, and greater capacity to highlight such considerations in trade-offs over water use and river management (see Jackson & O'Leary 2006). Empirical and theoretical studies of Indigenous resource use and socio-economic systems undertaken over many years by the Centre for Aboriginal Economic Policy Research (e.g. Altman 1987) provide a valuable basis for this field of inquiry. This work examined mechanisms to provide more resource rights to Indigenous

Australians, showing that the formation of new laws and new property rights is a significant trigger for better resource deals (Altman 2002).

More needs to be known about the size of the Indigenous agricultural sector and the nature of its demand for water, and the rules and norms likely to influence its valuation of options, preferences and trading choices. We do not know how changes to property rights may affect behaviour in Aboriginal societies, particularly with respect to land and water use decisions. The particularities of Indigenous property rights are not yet sufficiently understood nor accommodated in studies of market-based instruments, for example, the collective and indivisible nature of the title (Altman 2004a; Sheehan 2001; Langton 2006). Campbell (2002) discusses some of these theoretical and institutional issues in relation to Indigenous fisheries, arguing that ignorance of the economic characteristics of Indigenous rights hampers the valuation of benefits arising from those rights, as does the availability of appropriate measures of Indigenous value. We are aware of only a few economic valuations where techniques such as multi-attribute analysis (McDaniels & Trousdale 2005) and choice modelling (Windle & Rolfe 2002) have relied upon Indigenous input.

There is a lack of empirical data with which to properly evaluate the expectation or belief that the individual pursuit of self-interest will allocate water to the highest value and result in efficiencies in Indigenous contexts. There is also a poor understanding of the potential economic benefit to Indigenous people that might be realised from rights to water, and there are few insights into how to advance that understanding. Altman urges consideration of 'innovative approaches that might accord customary rights commercial (or quasi-commercial) status to encourage efficient use and possible trade' (2004a: 32; 2004b). The Native American water rights experience should be further evaluated for its relevance, as should the NSW approach to the development of an Aboriginal Water Trust.

Indigenous values, including cultural heritage, can be intangible and subjective, and hence difficult to relate explicitly to particular flow regimes and to quantify in allocation decisions (Jackson 2005; Douglas 2004). Indigenous groups that are unable to frame and specify their requirements may be distinctly disadvantaged when competing with organised groups with clearly articulated claims for water and recognised expectations for continuity of access and use. The above discussion reveals the need for greater clarity in conceptualising the types of Indigenous water uses or needs, as well as quantification and accounting techniques. Indigenous input to this conceptual deliberation will be crucial. Well-resourced cultural mapping exercises overseen by Aboriginal representative organisations show promise in documenting Indigenous values, and relating flow regimes to Indigenous values and places of significance. Support for this kind of basic research activity should be markedly increased to inform the water planning knowledge base and engage Indigenous people. Interdisciplinary environmental flows research that incorporates Indigenous knowledge into assessments and modelling is also required. Local studies are underway,[9] and US (Goodman 2000), New Zealand (Tipa & Teirney 2003) and South African methodologies (Pollard & Simanowitz 1997) may provide valuable insights into incorporating social values in flows research.

There is a pressing need to improve Indigenous people's knowledge of, and access to, water policy debates. It would be beneficial for Indigenous people to contribute to the formulation of benchmarks for Indigenous participation in water planning, for example. As there has been so little direct Indigenous contribution to date, and no national oversight and interventions since the demise of the Aboriginal and Torres Strait Islander Commission, this is a matter of some urgency. Altman (2004a) has also recommended a communication and public education strategy to explain native title rights to water. Development of guidelines to assist natural resource management groups to engage with Indigenous people and recognise their water interests would be equally valuable.

Conclusion

The Lingiari Foundation report on Indigenous water rights (2002) neatly summarised Indigenous aspirations for water use and management:

- to have their spiritual relationships to water resources protected
- to protect sites of significance
- to decisively participate in the better management of water resources
- to exercise their spiritual, cultural, social and economic rights through access to good-quality water and water for commercial use (McFarlane 2004: 10).

Indigenous interests were not then acknowledged by national water reform and Indigenous representatives took no part in contributing to the development of the NWI. The policy reform process has since partially responded to the Indigenous aspirations outlined above by recognising Indigenous needs to access water and to participate in water planning. Significant improvements in the knowledge base are required, particularly to clarify the nature of Indigenous requirements and develop consistent, practicable and credible methods for allocating water to Indigenous uses.

The NWI framework currently falls short in meeting the Indigenous economic aspirations referred to in this chapter. As a policy frontier creating new forms of property, it represents an opportunity to provide Indigenous access to commercially valuable water rights, arguably with fewer negative distributional effects than other established forms of property. Initiatives that align Indigenous access to water with economic development goals are consistent with international developments in Indigenous policy and could be further explored as states and territories seek to implement the NWI. As a matter of priority, attention should be given to engaging with Indigenous people, seeking their input to policy debates as well as their suggestions on implementation approaches. Indigenous forms of customary management and ecological knowledge can contribute to the serious challenge of sustainable water use, as they have in other areas of environmental policy such as biodiversity conservation. A high degree of Indigenous participation in new water management institutions and maintenance of customary management activity reliant on healthy water systems will benefit the entire nation.

References

Altman J (1987). *Hunter gatherers today: an Aboriginal economy in north Australia*. Australian Institute of Aboriginal Studies: Canberra.

Altman J (2002). The political economy of a treaty: opportunities and challenges for enhancing economic development for Indigenous Australians. *Drawing Board: An Australian Review of Public Affair* **3**, 65–81.

Altman J (2004a). Indigenous interests and water property rights. *Dialogue* **23**, 29–43.

Altman J (2004b). Economic development and Indigenous Australia: contestation over property, institutions and ideology. *Australian Journal of Agricultural and Resource Economics* **48**, 513–534.

Anon (2003). Report to the Murray-Darling Basin Commission: Indigenous response to the Living Murray Initiative. Murray Darling Basin Commission: Canberra.

Australian Heritage Commission (2002). *Ask first: a guide to respecting Indigenous heritage places and values*. Commonwealth of Australia: Canberra.

Baker R, Davies J & Young E (eds) (2001). Working on country: contemporary Indigenous management of Australia's lands and coastal regions. Oxford University Press: Oxford.

Barber K & Rumley H (2003). *Gunanurang (Kununurra). Big river, Aboriginal cultural values of the Ord River and wetlands*. Report prepared for the Western Australian Water and Rivers Commission: Perth.

Bartlett R (1997). Native title to water. In *Water law in Western Australia: comparative studies and options for reform* (eds R Bartlett, A Gardner & S Mascher), pp. 125–160. Centre of Commercial and Resources Law, University of Western Australia: Perth.

Behrendt J & Thompson P (2004). The recognition and protection of Aboriginal interests in New South Wales rivers. *Journal of Indigenous Policy* **3**, 37–140.

Bruns B & Meinzen-Dick R (2000). *Negotiating water rights.* Intermediate Technology: London.

Byrne D, Brayshaw H & Ireland T (2003). *Social significance.* Discussion paper. NSW National Parks Services Research Unit, Cultural Heritage Division: Sydney.

Campbell D (2002). The Indigenous sector: an economic perspective. In *Valuing fisheries: an economic framework* (ed. T Hundloe), pp. 166–186. University of Queensland Press: St Lucia.

Carlsson L & Berkes F (2005). Co-management: concepts and methodological implications. *Journal of Environmental Management* **75**, 65–76.

Cohen I (2004). Water Management Amendment Bill. Second Reading Speech, 23 June 2004. NSW Legislative Council Hansard, <http://www.parliament.nsw.gov.au>. Accessed 6 October 2006.

Collings N (2002). Water rights and international law. In *Background briefing papers: Indigenous rights to waters.* (ed. Lingiari Foundation), pp. 72–84. Lingiari Foundation: Broome.

Commonwealth of Australia (2004). *Guidelines for Indigenous participation in natural resource management.* Commonwealth of Australia: Canberra.

Connell D, Dovers S & Grafton R (2005). A critical analysis of the National Water Initiative. *Australasian Journal of Natural Resources Law and Policy* **10**, 81–107.

Delli Priscoli J (1998). Water and civilization: using history to reframe water policy debates and to build a new ecological realism. *Water Policy* **1**, 623–636.

Douglas P (2004). Healthy rivers and Indigenous interests. *Indigenous Law Bulletin* **5**, 12–14.

Edmunds M (ed.) (1998). *Regional agreements: key issues in Australia.* Native Title Unit, Australian Institute of Aboriginal and Torres Strait Islander Studies: Canberra.

English A (2002). More than archaeology: developing comprehensive approaches to Aboriginal heritage management in NSW. *Australian Journal of Environmental Management* **9**, 218–227.

Getches D (2005). Defending Indigenous water rights with the laws of the dominant culture; the case of the United States. In *Liquid relations: contested water rights and legal complexity* (eds D Roth, R Boelens & M Zwarteveen), pp. 44–65. Rutgers University Press: New Brunswick.

Gibbs M (2005). The right to development and Indigenous peoples: lessons from New Zealand. *World Development* **33**, 1365–1378.

Goldflam R (1997). Noble salvage: Aboriginal heritage protection and the Evatt Review. *Aboriginal Law Bulletin* **3**, 4–8.

Goodall H (2002). The river runs backwards. In *Words for country: landscape and language* (eds T Bonyhady & T Griffith), pp. 30–51. University of New South Wales Press: Sydney.

Goodman E (2000). Indian tribal sovereignty and water resources: watershed, ecosystems and tribal co-management. *Journal of Land, Resources and Environmental Law* **20**, 185–221.

Hancock N (1995). How to keep a secret. *Aboriginal Law Bulletin* **3**, 4–9.

Horton D (ed.) (1994). *Encyclopaedia of Aboriginal Australia: Aboriginal and Torres Strait Islander history, society and culture.* Aboriginal Studies Press for the Australian Institute of Aboriginal and Torres Strait Islander Studies: Canberra.

Howitt R (1993). Social impact assessment as 'applied people's geography'. *Australian Geographic Studies* **31**, 127–140.

Howitt R (2001). *Rethinking resource management: justice, sustainability and Indigenous peoples.* Routledge: London.

Jackson S (1997). A disturbing story: the fiction of rationality in land use planning in Aboriginal Australia. *Australian Planner* **34**, 15–18.

Jackson S (2004). *Preliminary report on Aboriginal perspectives on land-use and water management in the Daly River region, Northern Territory*. Report to the Northern Land Council, Darwin. CSIRO Sustainable Ecosystems. <http://www.terc.csiro.au>. Accessed 12 April 2007.

Jackson S (2005). Indigenous values and water resource management: a case study from the Northern Territory. *Australasian Journal of Environmental Management* **12**, 136–146.

Jackson S (2006). Compartmentalising culture: the articulation and consideration of Indigenous values in water resource management. *Australian Geographer* **37**, 19–32.

Jackson S & Langton M (2006). Indigenous cultural values and water resource management: workshop overview and introduction. In *Recognising and protecting Indigenous values in water resource management*. Report from a workshop held in Darwin, 5–6 April 2006 (ed S. Jackson). CSIRO: Berrimah. <http://www.terc.csiro.au>. Accessed 20 November 2006.

Jackson S & O'Leary P (2006). *Indigenous interests in tropical rivers: research and management issues*. Report to Land and Water Australia, North Australian Indigenous Land & Sea Management Alliance, Darwin. <http://www.terc.csiro.au>. Accessed 20 November 2006.

Jackson S, Storrs M & Morrison J (2005). Recognition of Aboriginal rights, interest and values in river research and management: perspectives from northern Australia. *Ecological Management and Restoration* **6**, 105–110.

Kaufmann P (ed.) (2004). *Water and fishing: Aboriginal rights in Australia and Canada*. Aboriginal and Torres Strait Islander Commission: Canberra.

Kolig E (1996). Thrilling the clay of our bodies: natural sites and the construction of sacredness in Australian Aboriginal and Austrian traditions, and in new age philosophy. *Anthropological Forum* **7**, 351–381.

Lane M (1997). Aboriginal participation in environmental planning. *Australian Geographical Studies* **35**, 308–323.

Lane M & Corbett T (2005). The tyranny of localism: Indigenous participation in community-based environmental management. *Journal of Environmental Policy and Planning* **7**, 141–159.

Langton M (1998). *Burning questions: emerging environmental issues for Indigenous peoples in northern Australia*. Centre for Indigenous Natural and Cultural Resource Management, Northern Territory University: Darwin.

Langton M (2002). Freshwater. In *Background briefing papers: Indigenous rights to waters* (ed. Lingiari Foundation), pp. 43–64. Lingiari Foundation: Broome.

Langton M (2006). Earth, wind, fire and water: the social and spiritual construction of water in Aboriginal societies. In *The Social Archaeology of Australian Indigenous Landscapes* (eds B David, B Barker & I McNiven), pp. 139–160. Aboriginal Studies Press: Canberra.

Langton M & Palmer L (2003). Modern agreement making and Indigenous people in Australia: issues and trends. *Australian Indigenous Law Reporter* **8**, 1–31.

Langton M, Tehan M, Palmer L & Shain K (eds) (2004). *Honour among nations? Treaties and agreements with Indigenous people*. Melbourne University Press: Melbourne.

Larsen L & Pannell S (2005). *Developing the wet tropics: Aboriginal cultural and natural resource management plan*. Report to Wet Tropics World Heritage Management Authority, Cairns. <http://jcu.edu.au/rainforest/publications/plan_proceedings–1.pdf>. Accessed 30 October 2006.

Lingiari Foundation (2002). *Onshore water rights. A discussion booklet for the Aboriginal and Torres Strait Islander Commission*, vol. 1. Lingiari Foundation: Broome.

Lombardi L (2004). American Indian water rights. *Indigenous Law Bulletin* **5**, 24–26.

Maddock, K (1982). *The Australian Aborigine: a portrait of their society*, 2nd edn. Penguin: Melbourne.

McDaniels T & Trousdale W (2005). Resource compensation and negotiation support in an aboriginal context: using community-based multi-attribute analysis to evaluate non-market losses. *Ecological Economics* **55**, 173–186.

McFarlane B (2004). *The National Water Initiative and acknowledging Indigenous interests in planning.* Paper presented to the National Water Conference, Sydney, 29 November 2004. <http://www.nntt.gov.au>. Accessed 15 August 2006.

McGurk B, Sinclair A & Diduck A (2006). An assessment of stakeholder advisory committees in forest management: case studies from Manitoba, Canada. *Society and Natural Resources* **19**, 809–826.

McKay J (2002). Onshore water project: briefing paper. In *Background briefing papers: Indigenous rights to waters* (ed. Lingiari Foundation), pp. 65–70. Lingiari Foundation: Broome.

Molle F (2004). Defining water rights: by prescription or negotiation? *Water Policy* **6**, 207–227.

Morgan M, Strelein L & Weir J (2004). Indigenous water rights within the Murray Darling Basin. *Indigenous Law Bulletin* **5**, 17–20.

Morgan M, Strelein L & Weir J. (in press). Authority, knowledge and values: Indigenous nations engagement in the management of natural resources in the Murray-Darling Basin. In *Settling with Indigenous peoples* (eds M Langton, O Mazel, L Palmer, K Shain & M Tehan). Federation Press: Sydney.

Neate G (2002). *Native title ten years on: getting on with the job or sitting on the fence.* Paper delivered to Native Title Update Forum, National Farmers Federation, Carnarvon, Western Australia, 21 May 2002. <http://www.nntt.gov.au>. Accessed 21 September 2006.

Neate G (2004). Agreement making and the Native Title Act. In *Honour among nations? Treaties and agreements with Indigenous people* (eds M Langton, M Tehan, L Palmer & K Shain), pp. 176–188. Melbourne University Press: Melbourne.

North Australian Indigenous Land & Sea Management Alliance (2006). *Living on saltwater country: review of literature about Aboriginal rights, use, management and interests in northern Australia marine environments.* Report to the National Oceans Office: Hobart.

NSW Department of Natural Resources (2006). *Water management plans: information for Aboriginal water users.* <http://www.dlwc.nsw.gov.au/water/info_aboriginal_water.shtml>. Accessed 26 September 2006.

NSW Healthy Rivers Commission (1999). *Independent inquiry into the Clarence River System.* Final Report, Healthy Rivers Commission of NSW, Sydney. <http://www.planning.nsw.gov.au/programservices/pdf/clarence_final.pdf>. Accessed 18 December 2006.

O'Connor R, Quartermain G & Bodney C (1989). *Report on an investigation into Aboriginal significance of wetlands and rivers in the Perth–Bunbury region.* Western Australian Water Resources Council: Perth.

O'Donnell M (2004). Briefing paper for the water rights project by the Lingiari Foundation and ATSIC. In *Background briefing papers: Indigenous rights to waters* (ed. Lingiari Foundation), pp. 95–105. Lingiari Foundation: Broome.

O'Faircheallaigh C & Corbett T (2005). Indigenous participation in environmental management of mining projects: the role of negotiated agreements. *Environmental Politics* **14**, 629–647.

Palmer K (1991). *Aborigines, values and the environment.* Fundamental Questions Paper No. 7, Centre for Resource and Environmental Studies, Australian National University: Canberra.

Paton S, Curtis A, McDonald G & Woods M (2004). Regional natural resource management: is it sustainable? *Australasian Journal of Environmental Management* **11**, 259–267.

Pollack D (2001). *Indigenous land in Australia: a quantitative assessment of Indigenous land holdings in 2000.* CAEPR Discussion Paper No. 221, Centre for Aboriginal Economic Policy Research, Australian National University: Canberra.

Pollard S & Simanowitz A (1997). Environmental flow requirements: a social dimension. In *Proceedings of the 23rd World Environment and Development Conference, Water and Sanitation for All: Partnerships and Innovations*, pp. 397–400. Durban, South Africa.
Rose B (1995). *Land management issues: attitudes and perceptions amongst Aboriginal people of central Australia*. Report to the Central Australia Land Council: Alice Springs.
Rose DB (1996). *Nourishing terrains: Australian Aboriginal views of landscape and wilderness*. Australian Heritage Commission: Canberra.
Rose DB (1999). Indigenous ecologies and an ethic of connection. In *Global ethics and environment* (ed. N Low), pp 175–187. Routledge: London.
Rose DB (2004). Freshwater rights and biophilia: Indigenous Australian perspectives. *Dialogue* **23**, 35–43.
Ross H (1992). Opportunities for Aboriginal participation in Australian social impact assessment. *Impact Assessment Bulletin* **10**, 47–76.
Roth D, Boelens R & Zwarteveen M (eds) (2005). *Liquid relations: contested water rights and legal complexity*. Rutgers University Press: New Brunswick.
Sharp N (2002). *Saltwater people: the waves of memory*. Allen & Unwin: Sydney.
Sheehan J (2001). Indigenous property rights and river management. *Water Science and Technology* **43**, 235–242.
Singh N (2006). Indigenous water management systems: interpreting symbolic dimensions in common property resource regimes. *Society and Natural Resources* **19**, 357–366.
Smith D (1998). *Indigenous land use agreements: the opportunities, challenges and policy implications of the amended Native Title Act*. Discussion Paper No. 163, Centre for Aboriginal Economic Policy Research, Australian National University: Canberra
Smyth D (2002). Appendix B: The Indigenous sector – an anthropological perspective. In *Valuing fisheries: an economic framework* (ed. T Hundloe), pp. 221–229. University of Queensland Press: St Lucia.
Smyth D, Szabo S & George M (2004). *Case studies in Indigenous engagement in natural resource management in Australia*. Report to the Commonwealth Department of Environment and Heritage: Canberra.
Storrs M, Yirbarbuk D, Whitehead P & Finlayson C (2001). The role of western science in contemporary Aboriginal community wetlands management in the Top End of Australia. In *Science and local communities: strengthening partnerships for effective wetland management* (eds M Carbonnell, N Nathai-Gyan & C Finlayson), pp. 7–13. Ducks Unlimited: US.
Strang V (2001). Poisoning the rainbow: mining, pollution and Indigenous cosmology in far north Queensland. In *Mining and Indigenous lifeworlds in Australia* (eds A Rumsey & J Weiner), pp. 208–225. Crawford House Publishers: Adelaide.
Strang V (2004). *The meaning of water*. Berg: Oxford.
Strelein L (2004). Symbolism and function: from native title to Aboriginal and Torres Strait Islander self-government. In *Honour among nations? Treaties and agreements with Indigenous people* (eds M Langton, M Tehan, L Palmer & K Shain), pp. 189–202. Melbourne University Press: Melbourne.
Tan P (1997). Native title and freshwater resources. In *Commercial implications of native title* (eds B Horrigan & S Young), pp. 157–190. Federation Press: Sydney.
Tipa G & Teirney L (2003). *A cultural health index for streams and waterways: indicators for recognising and expressing Maori value*. Report to the New Zealand Ministry for the Environment: Wellington.
Trigger D (1985). The social meaning of water for Aboriginal people in Australia's Gulf country. Unpublished paper delivered at School of Environmental Studies, Griffith University.

Tsamenyi M & Mfodwo K (2000). *Towards greater Indigenous participation in Australian commercial fisheries: some policy issues.* ATSIC National Policy Office Paper Series. Aboriginal and Torres Strait Islander Commission: Canberra.

Toussaint S, Sullivan P & Yu S (2005). Water ways in Aboriginal Australia: an interconnected analysis. *Anthropological Forum* **15**, 61–74.

Weiner J, Godwin L & L'Oste-Brown S (2002). *Australian Aboriginal heritage and native title: an example of contemporary Indigenous connection to country in central Queensland.* Occasional Paper Series, No. 1. National Native Title Tribunal: Perth.

Weir J (2006a). Cultural flows in the Murray Lower Darling Rivers. In *Recognising and protecting Indigenous values in water resource management.* Report from workshop held in Darwin, 5–6 April 2006 (ed. S Jackson), pp. 32–42. CSIRO: Berrimah. http://www.terc.csiro.au. Accessed 20 November 2006.

Weir J (2006b). Making the connection between water and sustaining Indigenous cultural life. In *People, practice and policy: a review of social and institutional research*, Social and Institutional Research Program online magazine, Land and Water Australia. www.sirp.gov.au/peoplepracticepolicy/a6.htm. Accessed 6 July 2006.

Whelan J & Oliver P (2005). Regional community-based planning: the challenge of participatory environmental governance. *Australasian Journal of Environmental Management* **12**, 126–135.

Willis E, Pearce M, Jenkin T, Wurst S & McCarthy C (2004). *Water supply and use in Aboriginal communities in South Australia.* Report to South Australian Department of Aboriginal Affairs and Reconciliation, Flinders University: Adelaide.

Windle J & Rolfe J (2002). *Natural resource management and the protection of Aboriginal cultural heritage.* Occasional Paper No. 5. Institute for Sustainable Regional Development, Central Queensland University: Rockhampton.

Woenne-Green S, Johnston R, Sultan R & Wallis A (1994). *Competing interests: Aboriginal participation in national parks and conservation reserves in Australia: a review.* Australian Conservation Foundation: Melbourne.

Yu S (2000). *Ngapa Kunangkul: living water.* Report on the Indigenous cultural values of groundwater in the La Grange sub-basin. Western Australian Water and Rivers Commission: Perth.

Endnotes

1 In 2000 it was estimated that close to 20% of the Australian landmass was Indigenous owned (Pollack 2001).

2 In NSW a native title holder means a person who holds native title rights pursuant to a determination under the *Native Title Act 1993* (Cth) (McKay 2002). As there are still so few determinations, this level of proof may continue to substantially limit the number of instances in which water is allocated for native title purposes.

3 The Aboriginal cultural licences were listed as a ministerial exemption to the embargo on further licences for the north coast rivers (Miller, pers comm.).

4 For example, in 1994 pumping for irrigation was drying up Ban Ban Springs, a site of considerable cultural significance over a wide area of central Queensland (Evatt 1996). The site was registered, but as the bore did not physically interfere *per se* with the actual area of the water reserve where the springs occurred, Queensland heritage legislation could not be relied upon.

5 Native title was recognised by the High Court in *Mabo (No. 2)*, when the majority of the Court held that 'the common law of this country recognises a form of native title which, in the cases where it has not been extinguished, reflects the entitlement of the Indigenous inhabitants, in accordance with their laws or customs, to their traditional lands' (cited in Neate 2004: 178). Soon after the *Mabo* decision of 1994, the *Native Title Act* was passed to provide certainty to existing title-holders and to

recognise and protect native title. Subsequent legal cases (e.g. *Ward* and *Yorta Yorta*) have confirmed the view that native title rights and interests 'must be construed as deriving from traditional law and custom. The common law recognises those rights and interests through the concept of native title' (Neate 2004: 191).

6 The term 'waters' has been defined in the *Native Title Act* to include 'sea, a river, a lake, a tidal inlet, a bay, an estuary, a harbour or subterranean waters; or the bed or subsoil under, or airspace over, any waters (including waters mentioned in paragraph (a)' (Tan 1997: 75).

7 According to McFarlane (2004), there were 600 native title claims Australia-wide by 2004 and most, if not all, water resources within the Australian mainland were subject to native title claims.

8 Commonwealth documents include *Guidelines for Indigenous Participation in Natural Resource Management* (Commonwealth of Australia 2004). Some states have also drafted Indigenous engagement guidelines.

9 The Daly River Fish and Flows project (led by Charles Darwin University) is collecting Indigenous fish knowledge and may use this in modelling environmental flows.

10 Commonwealth documents include Guidelines for Indigenous Participation in Natural Resource Management (Commonwealth of Australia 2004). Some states have also drafted Indigenous engagement guidelines.

Acknowledgments

Several people assisted in providing information for this chapter: Jessica Weir, David Miller, Christa Hay, Lesley Dias and Virginia Falk. We are most grateful to Jon Altman, Poh Ling Tan, Alan Andersen and Andra Putnis for their helpful comments on an early draft. We would also like to acknowledge the work of the conference organisers and editorial committee in producing this collection.

Chapter 4

Environmental water allocations and their governance

Alex Gardner and Kathleen H Bowmer

The Australian Water Resources 2005 baseline assessment of water resources at the beginning of the 2004 Intergovernmental Agreement on a National Water Initiative (NWI)[1] records in simple terms the law and policy for making environmental water allocations (EWAs)[2] and the level of EWAs that have been provided (National Water Commission 2006b). While there has been much progress in the provision of EWAs since the Council of Australian Governments agreed upon the Water Reform Framework Agreement in 1994 (1994 framework agreement),[3] the 2005 baseline assessment reveals great variation between jurisdictions in the expression of legal and policy provisions and in the extent of EWAs provided. Much remains to be done to achieve nationally the NWI objectives for EWAs (Gardner 2006), including:

- water provided to meet agreed EWA outcomes defined in water plans be given statutory recognition with at least the same degree of security as water access entitlements for consumptive use and be fully accounted for (COAG 2004a; NWI paras 23(iii), 25(ii), 35(i))
- all currently overallocated or overused systems be returned to environmentally sustainable levels of extraction (COAG 2004a; NWI paras 23(v), 41–43)[4]
- management practices and institutional arrangements identify the EWA outcomes as specifically as possible and establish accountable environmental water managers with the necessary authority and resources to achieve those outcomes (COAG 2004a; NWI para. 78).

Achieving these objectives raises various issues of governance, some of which were addressed in the National Water Commission's (Commission) 2005 National Competition Policy assessment of water reform progress (2005 Assessment Report) (National Water Commission 2006a). In this chapter, we primarily address some abiding concerns about the making of EWAs and comment briefly on two other issues; namely, the role of environmental water managers, and the nature of performance indicators used in monitoring, reporting on and evaluating the implementation of EWAs. We discuss what we know so far on each issue, what can be adapted from related resource governance experience to address the issue, and the research challenges.

1 Making EWAs: concerns with transparency and security

The Commission's 2005 Assessment Report identified the following concerns with the process of making EWAs, especially but not exclusively in relation to NSW (National Water Commission 2006a: 2.7& 2.28 (re NSW), 3.9 & 3.21–3.22 (Vic.), 4.8 & 4.21–4.22 (Ql), 6.23 (SA)):

- the adequacy of ecological science used to inform the preparation of water-sharing plans

- a lack of transparency in the trade-offs between consumptive and environmental water
- the consequent failure to deal with overallocation and the potential activation of sleeper and dozer entitlements if liberal trade rules are applied.

We would add the priority that is accorded to EWAs, though this issue seems to be characterised under the NWI as more one of legal security equivalent to that of water access entitlements. Before addressing these issues, we describe some relevant policy background.

1.1 Policy background to EWAs

> The goal of providing water for the environment is to sustain and where necessary restore ecological processes and biodiversity of water dependent ecosystems (ARMCANZ 1996: iii).

The common interpretation of the 1996 National Principles for the Provision of Water for Ecosystems was that water resource planning should, on the basis of the best available scientific information/knowledge,[5] provide water to sustain the ecological values of water-dependent ecosystems before water is allocated to human consumptive uses. In other words, water for the environment comes first. Where there are existing users, the provision of EWAs should go as far as possible towards sustaining the ecological values 'while recognising the existing rights of other water users' (ARMCANZ 1996: principle 4). Where the ecological water requirements could not be met because of existing uses, 'action (including reallocation) should be taken to meet environmental needs' (ARMCANZ 1996: principle 5). There was an expectation that, in due course, all water uses would be managed in a way that is ecologically sustainable (ARMCANZ 1996: principle 9).

Furthermore, the early common approach to the implementation of the 1994 framework agreement was that property rights in water would not be conferred 'until a comprehensive planning system is in place which fully describes the resource and establishes a framework of consumptive and environmental uses' (ARMCANZ 1995: executive summary). Water resource plans were to establish 'the environmental constraints within which any market for water entitlements might operate' (ARMCANZ 1995: 5). At the same time, it was anticipated that, with the exception of passing flow provisions, EWAs (at least for surface water) should be tradable within clear constraints to protect their essential environmental purpose (ARMCANZ 1995: executive summary).

A selective review of state legislation reveals varied implementation in law of this early common understanding of the goal of making EWAs. The legislative model enacted by NSW and SA appears to adopt the principle of priority for EWAs. The initial terms of the *Water Management Act 2000* (NSW) expressed priority for EWAs as a legal duty, though subsequent amendment of the Act has considerably weakened the nature of the legal duty (Gardner 2006: 215–218). Nevertheless, in its submissions to the NWC's 2005 assessment, NSW maintained that the *Water Management Act* 'gives priority to the supply and protection of water for the environment and for other public benefits' (National Water Commission 2006a: 2.8).

The *Natural Resources Management Act 2004* (SA) may achieve a similar effect. Its objects include 'to assist in the achievement of ecologically sustainable development in the State'; that concept is defined with a substantive content, albeit one that does not refer to water for the environment (section 7). The SA Act also provides that a water allocation plan must assess the quantity and quality of water needed by water-dependent ecosystems and provide for the allocation and use of water so that 'an equitable balance is achieved between environmental, social and economic needs for the water' (section 76(4)). The Commission says of the SA water allocation planning process that the 'catchment and the requirements of the associated ecosystems are assessed before the consumptive pool is determined for a system' (National Water Commission 2006a: 2.26 & 6.15). As with NSW, it is doubtful that the SA legislation provides a justiciable standard for testing the validity of an EWA in a water allocation plan except in extreme cases of imbalance (*Project Blue Sky Inc v Australian Broadcasting Authority*; Gardner 2006: 217). Nevertheless, the Commission was satisfied that SA's water management arrangements were consistent with its COAG commitments, but doubted that the practical implementation of NSW's arrangements were (National Water Commission 2006a: 2.26 & 6.15).

The contrasting legislative model of Victoria and Queensland confers more executive discretion. Victoria has legislated to establish an 'environmental water reserve ... to preserve the environmental values and health of water ecosystems,[6] but the content of the reserve is entirely a matter of ministerial discretion (Gardner 2006: 219–220). *The Water Management Act 2000* (Qld) provides for the goal of advancing the 'sustainable management' of water resources through the making of water resource plans (sections 11 & 38), but the plan's ecological outcomes and the provision of EWAs to meet them is entirely a matter of government discretion (sections 46, 47 & 50). Despite the contrasting legislative frameworks, the Commission declared its satisfaction that Victoria's water management arrangements are generally in line with the National Principles for the Provision of Water for Ecosystems (National Water Commission 2006a: 3.20). Similarly, it was satisfied that the Queensland arrangements for water management are consistent with COAG commitments and have legislative backing (National Water Commission 2006a: 4.20). The Commission's conclusions may reflect a new understanding of what is needed to meet the goal of making EWAs, and that new understanding is based on the NWI.

The NWI is more equivocal about the scope of the objective of ecological sustainability because it particularises the water resources subject to the goal. For example, one NWI action emphasises the need for legal security for water provided by the states to meet environmental outcomes '*as defined within relevant water plans*' (COAG 2004a; NWI para. 35, emphasis added). Thus, the environmental outcomes to be pursued are those defined in plans, and the plans' outcomes are determined in exercise of ministerial discretion rather than a legislative direction that gives priority to the environment.

Another major NWI objective is to 'complete the return of all currently overallocated or overused systems to *environmentally sustainable levels of extraction*' (COAG 2004a; NWI paras 5 & 23(iv)). That term is defined to mean the level of water extraction from a particular system which, if exceeded, would compromise key environmental assets or ecosystem functions and the productive base of the resource (COAG 2004a; NWI Schedule B1, emphasis added). The NWI further defines:

- 'overallocation' as situations where the total volume of water able to be extracted at a given time exceeds the environmentally sustainable level of extraction *for a particular system* (COAG 2004a; NWI Schedule B1, emphasis added)
- 'overuse' means where the total volume of water actually extracted *in a particular system* at a given time exceeds the environmentally sustainable levels of extraction for that system (COAG 2004a; NWI Schedule B1, emphasis added).

This objective and its associated definitions tend to suggest that there are clear criteria for ascertaining the water systems that need to be returned to environmentally sustainable levels of extraction. However, on further reading of the NWI, we see that overallocated/overused systems are those that were identified either by the 1999 agreement between the states and the National Competition Council, or by the states in accordance with a relevant water plan (COAG 2004a; NWI paras 41 & 43).

A state NWI implementation plan may identify which systems are regarded as overallocated or overused.[7] The NSW implementation plan acknowledges that the first group of water-sharing plans, which commenced in July 2004, address overuse 'in the context of the community's capacity to adjust to reductions in water allocations over the ten-year terms of these first plans' (NSW 2006, Action 5: 21). NSW asserts that the plans do this by establishing and managing long-term annual extraction limits, 'thereby protecting the proportion of river flows identified for fundamental ecosystem needs' (NSW 2006, Action 5: 21). NSW also says that the long-term average annual extraction limits enable it to manage the activation of sleeper and dozer licences, especially in overallocated systems. The Commission rejects this assertion for the reasons listed above (National Water Commission 2006a: 2.7).

1.2 Commission concerns with making EWAs

We consider here the Commission's reasons for questioning the transparency and security of the NSW EWAs, the knowledge we may adapt to addressing the issues and the research challenges presented.

First, the Commission found that NSW did not employ the best available science in preparing all the water-sharing plans (National Water Commission 2006a: 2.10–2.23). The ecological information was often too generic and not sufficiently detailed about specific catchments to enable the planning committees to determine the flow requirements needed to maintain ecosystem health (National Water Commission 2006a: 2.27).[8] While considerable effort was directed to generating information about the potential socio-economic impacts of the proposed water-sharing plans, there was not the same endeavour in researching the effects of various options for EWAs. Although environmental scientists recommended an increase in the amount of water provided for the environment, the reduction of the long-term annual average volume of extractions was 'arbitrarily limited to 10 per cent as a result of the environmental rules imposed by the regulated river water-sharing plans' (National Water Commission 2006a: 2.18–2.24).

Does using the 'best available scientific knowledge', as the requirement is expressed in the NWI (COAG 2004a; NWI Schedule E, para. 6(iii)), entail new studies to gather new information in order to ascertain the ecological requirements of the particular water system in preparing the plan? To answer this question, we should initially review the terms of the relevant legislation to see what it says about the process of gathering and considering information in the preparation of the plan. That is not an exercise to be conducted here. However, this general issue was considered in the context of forestry management in *Bridgetown/Greenbushes Friends of the Forest v Executive Director of Department of Conservation and Land Management*. The plaintiff forest conservation groups argued that the statutory duty on the defendant to manage state forests in accordance with the regional management plan required that the defendant fulfil its plan commitment to consult with the community and to consider relevant factors under the plan, by undertaking adequate studies of the areas of forest proposed to be logged before approving the logging plans. In essence, it was argued that any decision to approve the logging plans should be based on adequate information. It was held that 'the decision-maker is not required to gather or assemble information which is not readily available'. Applying that proposition to the pleadings in the case, it was held that the plaintiffs had failed to establish that 'the information relied on is readily accessible' (per Templeman J at 419–431; Scott J agreed with Templeman J). Indeed the requirement, as pleaded, to undertake scientific studies to gather the information meant that it was not readily available.

The gathering and consideration of information are processes of general importance in all resources management and environmental regulation. The legal sufficiency of how those processes are conducted seems to have been considered most in relation to environmental impact assessment, a special tool for managing the adverse impacts of resource use on the environment. There is much legal learning on the topic (Bates 2006: esp. at 358–361; MacLennan & Gardner 1997). The common test of the 'adequacy' of an environmental impact statement is that it must be sufficiently specific to direct a reasonably intelligent and informed mind to the possible environmental consequences of the proposed development (*Schaffer Corporation Ltd v Hawkesbury City Council* (p. 31)). It may be that this sort of proposition should be applied to the scientific information to be assembled in the preparation of a water-sharing plan that determines an EWA, and that this test will require undertaking new research if the existing information is not specific enough to understand the likely environmental consequences.

In addition to the problems related to the use of best available science, the Commission also found that there was a lack of transparency in the process of making the trade-offs between the environmental and consumptive allocations. There were problems in the process of public consultation in relation to the timeliness of information provided to stakeholders, the type and quality of that information, and the transparency of the process by which stakeholders could participate because there was a lack of opportunity to test advice provided by government officers in presentations to planning committees (National Water Commission 2006a: 2.18–2.27). NSW stated that the planning committees saw 'their role in the water-sharing planning process [as] to establish through legislation a legal share of water for the environment, and that the actual volume of allocation provided for the environment could be increased in the future' (National Water Commission 2006a: 2.19).

What are the legal standards for consultation and transparency in decision-making? Again, the legal starting-point is the interpretation of the relevant statute, but again that is not an exercise to be conducted here. However, it is worth noting that the preparation of the first group of water-sharing plans in NSW was undertaken in slightly unusual circumstances that may explain some of the subsequent discontent with the process. All the initial plans were made as 'minister's plans' rather than as plans prepared by management committees for the approval of the minister (*Water Management Act 2000* (NSW) Parts 2, 3 & 4).[9] This led to the argument in *Murrumbidgee Groundwater Preservation Association v the Minister for Natural Resources* that the minister failed to follow the standard statutory procedures in order to avoid the standard consultation procedures, and that the minister's plans were invalid because they were made for an extraneous purpose – a purpose outside the scope of the statute. It was held that there was no evidence to support the argument because the minister had appointed consultative committees under the *Water Act 1912* (NSW). Surprisingly, the court also found that there was nothing in the scheme of the *Water Management Act 2000* to suggest that the minister's plans are a subordinate form of making a plan and one only to be exercised as a matter of last resort. This finding is even more incredible when considered in light of the standard procedures at that time, which authorised the minister to take over preparation of a plan if the management committee failed to prepare a draft plan in accordance with the terms of reference (section 15(3)). It is notable that this provision was amended in 2004 to empower the minister to prepare a minister's plan in these circumstances.

Besides the statutory procedures for consultation, the common law also supplies the rules of procedural fairness that require a decision-maker to consult a person or persons whose interests will be affected by the exercise of the power. Historically, there was some judicial reticence to apply these rules to executive decisions of a quasi-legislative or subordinate legislative character, such as making a plan, because natural justice protects the interests of individuals and instruments of a legislative character can affect many people and be made for broad policy reasons (Creyke & McMillan 2005L para. 7.4.4.). However, that reticence has diminished in the Australian context (*Bread Manufacturers of New South Wales v Evans* (per

Mason and Wilson JJ at 432), and there are now challenges to the validity of planning decisions on the basis of arguments of procedural fairness.

Such a challenge was made in *Upper Namoi Water Users Association Inc & Ors v Minister for Natural Resources*, but the decision of the NSW Land and Environment Court in that case related only to issues of court process. The substantive arguments on procedural fairness were left unresolved. The applicant's allegations included that the minister's determination to make a minister's plan rather than apply the planning procedures of Parts 2 and 3 of the *Water Management Act* involved a breach of procedural fairness and that there was a legitimate expectation that the plan would be prepared by a management committee and be made as a management plan rather than as a minister's plan. Bignold J refused the minister's request to hear a separate trial of the abstract questions of whether rules of procedural fairness applied to the statutory provisions governing management plans and minister's plans. His Honour said: 'even the proper construction of the relevant provisions of the *Water Management Act* to determine whether the rules of procedural fairness have been "excluded by plain words of necessary intendment" ... cannot be undertaken without a proper understanding of (i) the content of the Water Sharing Plan and (ii) the nature of the rights or legitimate expectations held by the various Applicants in the present case' (p. [20]). The case did not proceed to a decision on the substantive arguments.

There is clearly scope for further study of the appropriate procedural standards for consultation and transparency in the preparation and making of plans. At least, a more detailed review of the NSW process may help explain why the judicial review of the NSW process arrived at different conclusions from the Commission. It would also be instructive to review the law and commentary regarding the procedures for making land use plans (O'Mara 2004), a study that may provide valuable comparative lessons, including for legislative amendment.

A third failing identified by the Commission was that the problems in relation to the use of best available science and public participation produced a failure by the NSW government to deal with overallocation and the potential activation of sleeper and dozer entitlements if liberal trade rules are applied. This concern was raised by the Combined Environmental NGOs (National Water Commission 2006a: 2.6–2.7). The Commission concluded that, because of its concerns about the adequacy of the scientific information and the transparency of the process for making trade-offs between environmental and consumptive entitlements, 'it is difficult in many systems to conclude that the environmental allocations are within a range of outcomes that could reasonably be reached on consideration of the best available science and robust socio-economic evidence' (National Water Commission 2006a: 2.23). In some systems, notably the Murrumbidgee Regulated River water-sharing plan, the level of overallocation has resulted in rules that prohibit temporary trade of high-security licence (annual) allocations after 1 September and of general-security licence allocations after the end of February (National Water Commission 2006a: 2.35). While the Commission acknowledged NSW's argument that these rules were intended to ensure that the environmental objective of the plan could be met, it noted (but only in very general terms) that there may be more adaptive approaches that give greater flexibility to the trading regime while also securing the environmental outcomes of the plan.

How do we define overallocation, and is it really a problem? Also, does it matter that a plan may restrict the right to trade water in order to prevent a problem of overallocation leading to a problem of overuse? The Murrumbidgee plan's constraint on trade criticised by the Commission has, in fact, been held valid by the NSW Land and Environment Court in *Murrumbidgee Horticulture Council v Minister for Land and Water (NSW)*. The source of the problem here[10] is mired in the historical practices of water allocation that gave horticulturalists with perennial crops generous rights to large volumes of high-security water and gave annual croppers large entitlements to general-security water. Historically, seasonal conditions meant that high-security licensees frequently did not need to take their full entitlements, and this permitted a higher reliability of allocations to the general-security licensees. Thus, the levels of diversion or use by the high-security licensees have been much lower than their levels of entitlement and annual allocation. Murrumbidgee Irrigation Ltd now holds entitlements to approximately 280 GL of high-security water and 930 GL of general-security water. If the high-security licensees were permitted to trade their full allocations at any time of the year, there would most likely be a greater than historical use of high-security water that, in combination with the historical levels of use of general-security water, would lead to an exceedance of the Murray-Darling Basin Commission cap and an inadequate provision of water for the EWA. The constraints on trade are therefore designed to preserve historical levels of water use, especially by high-security licensees. This has the effect of providing greater security to the EWA and greater reliability to general-security licensees without the dollar costs of trade (i.e. buying the water from high-security licensees). The Land and Environment Court upheld the constraints on trade because the *Water Management Act* gives no unfettered right to trade water.

This case also identifies another issue that is not explicitly dealt with in either the NWI or the Commission's 2005 Assessment Report; that is, the basis on which the old water licences under the *Water Act 1912* (NSW) were converted to the new tradable entitlements under the *Water Management Act*. It is common knowledge that the NSW government chose to honour 'legal entitlement' rather than 'history of use' to calculate the new entitlements. Apparently, this was not because there was a lack of data to determine the historical use, which may be a problem in some jurisdictions. Rather, the legal entitlement approach was regarded as more equitable because the sleeper and dozer licences (unused or partially used licences) were regarded as having a market value and were already being traded (Simpson 2006).[11] The policy of honouring sleepers, which has also applied in Victoria,[12] has been exacerbated by the further policy of permitting sleepers to be traded (Bond & Farrier 1996: 217; Pye 2006: 137–138). It appears that much of the water trade has been in sleeper allocations; CSIRO found in 2000 that 99% of water traded in the interstate pilot trade project came from sleepers (CSIRO Land & Water 2000: executive summary).[13] While the full extent of this practice may not have been publicly documented, the activation of even some sleeper allocations has surely exacerbated problems of overallocation and overuse.

The policy of honouring sleepers was also initially applied to groundwater in NSW (*Murrumbidgee Groundwater Preservation Association v the Minister for Natural Resources*). However, it has been revised in the 2006 amendment of the water-sharing plans for the six major inland alluvial groundwater resources of NSW, where a greater weighting has been given to historical use.[14] Professor Cullen, a National Water Commissioner, has stated (Cullen 2006: 7) 'Management agencies need to withdraw groundwater licences that have not been used. The failure to withdraw sleeper licences when surface water trading commenced saw them activated and traded with serious consequences.' However, withdrawing unused water from legal entitlements is questioned on the basis that it may disadvantage the licensees who have invested in efficiency and saved water, and reward those who use water inefficiently. It is arguable that water that is inefficiently used should be just as susceptible to withdrawal as water that is unused.

How do we respond to the problems of overallocation and overuse, especially from the activation of sleepers? With great respect to the authors of the Commission's 2005 Assessment Report, any suggestion that the constraint on trade in the Murrumbidgee plan should be removed must be preceded by a review of the approach to overallocation. On one view, overallocation can be controlled by limiting the annual allocations to ensure that the rules-based EWAs[15] are protected (Simpson 2006). This may reduce the reliability of entitlements, but is that a problem if the market factors it into the value of entitlements? If reduced allocations are not acceptable, then constraining the trade of sleeper allocations may be the only way to protect the rules-based EWAs and the levels of reliability of lower-security entitlements.

Consideration of these issues may involve two sources of comparative knowledge: the 'beneficial use' doctrine and the allocation of tradable entitlements in Australian fisheries law (see also Chapter 6, this volume). In water resources law, the beneficial use doctrine – the 'use it or lose it' policy – originated in the US doctrine of prior appropriation by which the appropriators of natural waters gained, for little or no cost, a priority right according to the initial time of appropriating the water to a beneficial use. This right of appropriation is a proprietary right, and tradable. In times of scarcity, those lower on the hierarchy of rights obtain water by trading up the hierarchy. While some commentators suggest that water trading commenced in Australia in response to increasing scarcity (Sturgess & Wright 1993: 12), it is possible that the operation of water markets in US prior appropriation jurisdictions fostered the Australian fascination with the establishment of water markets. Certainly, it was a similar market ethos that informed the argument of the Murrumbidgee Horticulture Council in challenging the water-sharing plan's constraint on its capacity to sell high-security water.

However, we should be wary of adopting or adapting other jurisdictions' doctrines without fully understanding them. The US prior appropriation right (Sax et al. 2000: ch. 3) applies only when water is put to a beneficial use, and continues on that condition too. A water rights owner who does not use their water for five years, or does not use it efficiently, may lose it. It has even been argued that inefficient use (in the form of a water utility not recycling effluent) should be a barrier to the acquisition of a new right of appropriation (Sax et al. 2000: 159 ff; Jackson 2005).[16] Importantly, the prevalent rule in US prior appropriation states is that only water rights that have historically been used may be transferred. On the other hand, it has been argued that the beneficial use doctrine should not be applied to 'anticipatory rights' (for proposed future use) acquired at auction (Sax et al. 2000: 258 ff; Williams 1983). This is certainly an important question in WA where a statutory-based 'use it or lose it' policy, which also precludes the transfer of sleeper licences, has been in spasmodic operation for some years (*Rights in Water and Irrigation Act 1914* (WA), Schedule 1, clause 24(2)(d); Water & Rivers Commission 2001, 2003). Some water reform advocates now suggest abandoning it in the move to fully tradable entitlements (Water Reform Implementation Committee 2006: 23–25).

A comparative study of the US beneficial use doctrine would be a valuable foundation on which to revisit the Australian approaches to determining the content and transferability of the new tradable water entitlements, particularly whether they should be on the basis of legal entitlement or historical use. Various research challenges will accompany this task. The operation of a beneficial use doctrine in the Australian context would have to be defined with various existing policy parameters in mind:

- whether the initial acquisition of an access entitlement was from a government grant and free of cost or was bought at auction
- the effect of borrow and carryover provisions which are designed to mitigate fluctuations in annual allocations and reduce the incentives for wasteful end-of-season use
- the distinction between permanent trade of access entitlements and temporary trade of annual allocations. In particular, where an access entitlement contains an element of 'insurance' against a drought year, should the trade of an annual allocation only be an amount of water equal to the long-term average annual use and not the full allocation
- what is an appropriate 'long term' to gauge average annual use?

There may also be some value in a comparative study of the allocation of fisheries quotas in Australia, where determinations of annual quotas have been made on basis of historical catch over a defined period of years (*Skoljarev v Australian Fisheries Management and Howard v Australian Fisheries Management Authority*). The fisheries case law reveals that there may be complex issues associated with calculation of historical use, including the consideration of personal circumstances that may warrant departure from a purely historical record of use.

1.3 Conclusion on concerns with making EWAs

It may be debatable whether inadequate science and non-transparent process in making water plans have been a cause or consequence of a lack of political will to address the problems of overallocation and overuse as a prelude to the initiation of full regimes of tradable water entitlements. Whatever our opinion on that question, it appears that the strategy for addressing these problems has reoriented around using the water market itself to recover water for the environment. An important part of this reoriented strategy is the use of public funds to acquire environmental water by sponsoring efficiency gains in water infrastructure or by the outright purchase of water access entitlements (COAG 2004a; NWI para. 79(ii)); COAG 2004b, para. 23).[17] As a consequence, the role of environmental water managers is becoming very important, not only in managing EWAs provided for in water plans, but in the acquisition of environmental water entitlements. As the Productivity Commission has recently commented (Productivity Commission 2006, p 151):

> Regulations and planning processes may underprovide for environmental needs in the hope or expectation that the environmental manager will provide the additional services to meet the desired environmental standard.

2 Environmental water managers

In the NWI, the states and territories agree to establish and equip 'accountable environmental water managers with the necessary authority and resources to provide sufficient water' for EWAs, including an authority to trade environmental water (COAG 2004a; NWI paras 78 & 79). 'Environmental [water] manager' is defined as an expertise-based function with responsibility for management of environmental water giving effect to the environmental water objectives of water plans. The environmental water may be based on plan rules or access entitlements. The institutional form of the environmental manager may vary from place to place; it may be a separate body or an existing basin, catchment or river manager, 'provided that the function is assigned the necessary powers and resources, potential conflicts are minimised, and lines of accountability are clear' (COAG 2004a; NWI Schedule B(i)).

The application of these propositions to state arrangements was considered by the Commission in its 2005 Assessment Report (National Water Commission 2006a: 2.60–2.63 (NSW), 3.45– .48 (Vic.), 4.45–4.47 (Qld), 6.46–6.49 (SA)). The Productivity Commission has also recently reviewed the role of environmental water managers (Productivity Commission 2006: 150–157). While there may be an array of environmental water managers at the local, regional, state and national levels, the primary model so far for environmental water managers in NSW, Victoria and SA is the catchment management authorities/natural resources management boards. In Queensland, where there are no tradable environmental water entitlements, it is the state Department of Natural Resources, Mines and Water. Various issues such as funding, independence and accountability are considered in these two reports. While these issues warrant more consideration, we make no particular suggestions for research directions other than the need to consider what organisations may assume the primary role of environmental water managers in states like WA and Queensland where there is no statutory foundation for regional natural resources management bodies.

All states and territories have a policy gap that warrants research – the suggestion that there should be a greater role for environmental non-government organisations (NGOs) in managing EWAs. This was the joint submission from Environment Victoria and the Australian Conservation Foundation to the Commission for improving the management of the environmental water reserve in Victoria (National Water Commission 2006a: 3.47). A similar proposal could be made for Indigenous peoples' organisations having the right to manage allocations to Indigenous interests that are not the subject of native title rights, which would be directly enforceable like any other access entitlement. It may be thought that the NWI agreement to establish and equip environmental water managers contemplates only public bodies performing these functions, but that need not be so.

Environmental NGOs already have the legal capacity to take up some of the roles of environmental water managers. In NSW and SA any person, including an NGO, can hold access entitlements for EWAs (*Water Management Act 2000* (NSW), section 8B; *Natural Resources Management Act 2004* (SA), section 146; National Water Commission 2006a: 6.13; Productivity Commission 2006: 195, Box 8.3).[18] In Victoria environmental entitlements are vested only in the Minister for Environment and may be managed by catchment management authorities (*Water Act 1989* (Vic.), Part 4, Div. 1A, sections 48B ff, 189). Under the intergovernmental agreement on Murray-Darling Basin environmental water, it would appear that NGOs may propose, invest in or implement 'eligible measures' for the recovery of environmental water and be responsible for local management of EWAs (COAG 2004b: paras 75 & 76). The framework for NGO involvement in managing EWAs is an important question, as is the issue of whether access entitlements acquired by environmental NGOs must be part of, or may be additional to, water plan provisions for EWAs.

Australian approaches to these issues may benefit from studying the legal framework of private water trusts in the US (King 2004). In Washington and Oregon, NGOs may acquire water rights for environmental purposes through sale, donation and lease, but they are prohibited from holding such rights – the EWAs must be transferred to the states. In particular, these NGOs have assisted in rule formation, water right acquisition, and monitoring and enforcement.

3 Monitoring and reporting EWAs: the nature of indicators

Environmental water accounting is an important aspect of the NWI. The parties agreed to develop a register of new and existing environmental water showing all relevant details, including the environmental outcomes sought, and to report annually on the extent to which environmental water rules 'were implemented and overall effectiveness of the use of resources in the context of environmental and other public benefit outcomes sought and achieved' (COAG 2004a; NWI para. 85). Two important legal questions arise. The first, the nature of the duty on environmental water managers to perform the monitoring and reporting functions, has been discussed elsewhere (Gardner 2006: 229). The statement and fulfilment of those duties needs further research in light of the Commission's concern that the NSW performance-monitoring programs had not been implemented for the water-sharing plans that commenced on 1 July 2004 (National Water Commission 2006a: 2.63). The second issue is the nature of the performance indicators expressed in water plans for monitoring and reporting on the implementation of EWAs. This requires some explanation.

How well are water plans specifying the environmental outcomes of EWAs? The Commission 2005 Assessment Report identified problems with the indicators expressed in the SA Clare Valley water allocation plan:

> The description of environmental water requirements is qualitative and generalised. The environmental flow needs are expressed very simplistically ... and the plan does not seem to quantify them in a way that can be used to assess clearly whether the water needs of the system are met or not (National Water Commission 2006a: 6.11–6.12).

While Victoria claims that it is monitoring against 'clear ecological objectives', the Commission concludes that 'it is not evident whether [the FLOWS methodology] includes an ongoing monitoring phase that targets key ecological and physical performance indicators tied to adaptive management' (National Water Commission 2006a: 3.13 & 3.12). In contrast, the Commission's comments about NSW are potentially comforting:

> The Commission is aware that the [NSW Act] requires that water-sharing plans include a series of indicators and targets to measure their environmental performance (National Water Commission 2006a: 2.63, emphasis added).[19]

However, the Commission understanding of the effect of performances indicators as targets is contrary to the opinion of the NSW Court of Appeal in the *Gwydir River Case* (*Nature Conservation Council of NSW Inc v Minister Administering the Water Management Act* (pp. 32–52)). In that case, the Nature Conservation Council argued that a performance indicator must include 'some quantity or quality that has been fixed or approved in the management plan, so that a person who applies the performance indicator is capable of ascertaining the result'. It argued that the indicators adopted by the plan, such as 'change in ecological condition of this water source and dependent ecosystems', failed to comply and that this affected the validity of the plan. The court rejected the argument, holding that the performance indicators did not need to set standards or targets; it was sufficient that they were expressed in general terms in the plan with supplementary guidance in an appendix (not part of the plan) as to the types of factors to be measured, but still with no standards or targets.

Undoubtedly, it would be difficult to agree on performance indicators that are targets for measuring the success of EWAs. This would be especially true if there were a lack of specific ecological studies to gather the baseline data on the current ecological conditions and those that should be achieved under the plan. But is that what the NWI requires us to do? Is that what we need to do if we are to achieve the goal of providing water for the environment, to sustain and where necessary restore ecological processes and biodiversity of water-dependent ecosystems? The answer requires careful research, not only on the legal propositions that may be brought to bear on the implementation of water plans. It will require a major exercise in scientific research to ascertain performance indicators and monitoring systems that gain sufficient community and political acceptance for them to be expressed in legally binding documents and enforced by the courts.

Conclusion

We have raised three policy gaps in the framework for the governance of EWAs, which are also identified in the Commission's 2005 Assessment Report. We have particularly discussed three abiding concerns with the process of making EWAs: the adequacy of scientific knowledge, transparency of process, and a failure to address issues of overallocation and overuse. We have recommended various lines of comparative study to address these concerns, but note that a significant research challenge in that line of inquiry will be to understand the variability in legislation for different resource management regimes.

There may be other research issues that have particular ramifications for the governance of EWAs:

- uncertainties underpinning the water balance, especially the interaction between surface and groundwater and the relationship between surface water run-off and prolonged drought and landscape change
- the prioritisation of EWAs for icon sites versus maintaining or improving general river function
- managing EWAs over a longer cycle than seasonal or annual priorities.

It is a constant challenge to comprehend the level of impact that particular practical issues can have on the design of a fundamental framework for governance.

The discussion in this chapter has also indicated another fundamental research and policy challenge for addressing the policy gaps in the framework for governance of EWAs – the interaction of EWAs with the framework for sharing scarcity between access entitlement holders. Australia has a tradition of sharing scarcity proportionally. In times of scarcity, reductions in entitlements have been shared equally though subject to the recognition of the special needs of different types of water users, e.g. a higher security of access entitlement was given to horticulturalists than to annual croppers. But this was a security to use the water, not to trade. With the reform of the legal framework for water management embracing tradable property rights to water, there is an evolving hierarchy of tradable entitlements. However, conversion of high-security licences to fully tradable entitlements risks converting an insurance against need into a profit from other people's needs in times of scarcity. This is a significant change in philosophy and marks a much greater reliance on the market to distribute resources in times of special need.

What will be the impact of this on EWAs? They are essentially public interests. Greater recourse to market distribution of resources reveals a tension between public regulatory models with greater use of public policy process and consultation to achieve public policy goals, and market-oriented models with greater use of private transactions that traditionally eschew public scrutiny. A very significant policy and research challenge is how to maintain public scrutiny of, and confidence in, market processes and results.

References

ARMCANZ (1995). *Water allocations and entitlements: a national framework for the implementation of property rights in water*. Agriculture and Resource Management Council of Australia and New Zealand, Standing Committee on Agriculture and Resource Management.

ARMCANZ (1996). *National principles for the provision of water for ecosystems*. Agriculture and Resource Management Council of Australia and New Zealand and Australian and New Zealand Environment and Conservation Council.

Bates G (2006). *Environmental law in Australia*, 6th edn. LexisNexis Butterworths: Sydney.

Bond M & Farrier D (1996). Transferable water allocations: property right or shimmering mirage? *Environmental and Planning Law Journal* **13**, 213.

Bowmer K (2006). Personal communication 31 October 2006. Professor Kathleen Bowmer, Charles Sturt University: NSW.

Council of Australian Governments (2004a). *Intergovernmental agreement on a national water initiative*. Available on Council of Australian Governments website: <http://www.coag.gov.au/meetings/250604/>.

Council of Australian Governments (2004b). *Intergovernmental agreement on addressing water overallocation and achieving environmental objectives in the Murray–Darling Basin*, 25 June 2004. Available on Council of Australian Governments website: <http://www.coag.gov.au/meetings/250604/index.htm>. Accessed on 13 November 2006.

Creyke R & McMillan J (2005). *Control of government action*. LexisNexis Butterworths: Sydney.

CSIRO Land and Water (2000). *Inter-state water trading: a two year review*, Executive Summary, December 2000. Available at <http://www.clw.csiro.au/publications/consultancy/2000/inter_trading.pdf>. Accessed on 11 November 2006.

Cullen P (2006). Flying blind: the disconnect between groundwater and policy. In *Proceedings of the 10th Murray–Darling Groundwater Workshop*, Canberra, 7 September 2006, p. 7. Available on National Water Commission website: <http://www.nwc.gov.au/news/speeches.cfm>.

Eberbach P (2006). Personal communication to Professor Kathleen Bowmer, October 2006. Associate Professor Phil Eberbach, Charles Sturt University, Wagga Wagga, New South Wales; one-time chairperson of the Lower Murray Groundwater Sharing Committee.

Gardner A (2006). Environmental water allocations in Australia. *Environmental and Planning Law Journal* **23**, 208.

Jackson H (2005). Potential exposure to legal liabilities for the supply of recycled water and biosolids. *Environmental and Planning Law Journal* **22**, 418.

King MA (2004). Getting our feet wet: an introduction to water trusts. *Harvard Environmental Law Review* **28**, 495.

MacLennan T & Gardner A (1997). The adequacy of information for environmental decision-making. *Australian Environmental Law News* **4**, 36–44.

National Water Commission (2006a). *2005 National competition policy assessment of water reform progress*, April 2006. Available at <http://www.nwc.gov.au/nwi/ncp_water_reform.cfm>.

National Water Commission (2006b). *Australian water resources 2005*, published October 2006. Available at <http://www.water.gov.au/Default.aspx>.

New South Wales (2006). Implementation plan for the National Water Initiative 2006. Available at National Water Commission website: <http://www.nwc.gov.au/nwi/nwi_implementation.cfm>.

O'Mara A (2004). Procedural fairness and public participation in planning. *Environmental and Planning Law Journal* **21**, 62.

Productivity Commission (2006). *Rural water use and the environment: the role of market mechanisms*, 11 August 2006, p. 151. Available at <www.pc.gov.au>.

Pye A (2006). Water trading along the Murray: a South Australian perspective. *Environmental and Planning Law Journal* **23**, 131.

Sax J et al. (2000). *Legal control of water resources: cases and materials*, 3rd edn. West Group: St Paul, Minnesota.

Simpson P (2006). Personal communication, 24 October 2006. Paul Simpson, Department of Natural Resources and Environment, NSW.

Sturgess G & Wright M (1993). *Water rights in rural New South Wales: the evolution of a property rights system*. Centre for Independent Studies.

Water and Rivers Commission (2001). *Statewide policy No.6: transferable (tradable) water entitlements for Western Australia*, October. Available at Department of Water website: <http://portal.water.wa.gov.au/portal/page/portal/dow/policy>. Accessed 11 November 2006.

Water and Rivers Commission (2003). *State policy No. 11: management of unused licensed water entitlements*, November. Available at Department of Water website: <http://portal.water.wa.gov.au/portal/page/portal/dow/policy>. Accessed 11 November 2006.

Water Reform Implementation Committee (2006). *A draft blueprint for water reform in Western Australia*, 25 July.

Williams S (1983). The requirement of beneficial use as a cause of waste in water resource development. *Natural Resource Journal* **23**, 7.

Cases

Bread Manufacturers of New South Wales v Evans (1981) 180 CLR 404.

Bridgetown/Greenbushes Friends of the Forest v Executive Director of Department of Conservation and Land Management (1997) 94 LGERA 380.

Howard v Australian Fisheries Management Authority [2006] FMCA 975.

Murrumbidgee Horticulture Council v Minister for Land and Water (NSW) [2003] NSW LEC 213; 127 LGERA 450.

Murrumbidgee Groundwater Preservation Association v Minister for Natural Resources [2005] NSWCA 10; 138 LGERA 11.

Nature Conservation Council of NSW Inc v Minister Administering the Water Management Act (2005) 137 LGERA 320; [2005] NSWCA 9.

Project Blue Sky Inc v Australian Broadcasting Authority (1998) 194 CLR 355.

Schaffer Corporation Ltd v Hawkesbury City Council (1992) 77 LGRA 21.

Skoljarev v Australian Fisheries Management Authority (1995) 39 ALD 517.

Upper Namoi Water Users Association Inc & Ors v Minister for Natural Resources [2003] NSWLEC 175.

Endnotes

1. A copy of the NWI is available on the Council of Australian Governments website: <http://www.coag.gov.au/meetings/250604/> and on the National Water Commission website: <http://www.nwc.gov.au/NWI/index.cfm>.
2. EWAs is here used to mean 'environmental and other public benefit outcomes' defined in water plans, as defined in the NWI in Schedule B(i), Glossary of Terms. EWAs may be allocated to the environment by plan rules or as access entitlements: NWI para. 35(ii).
3. COAG Communique (February 1994), Attachment A, Water Resource Policy. A copy of Attachment A can be found on the National Competition Council website, in the Compendium of National Competition Policy Agreements (1998), 2nd edn, p. 103: http://www.ncc.gov.au/publication.asp?publicationID=99&activityID=39.
4. The term 'environmentally sustainable levels of extraction' is defined in the NWI Schedule B(i) Glossary of Terms as 'the level of water extraction from a particular system which, if exceeded, would compromise key environmental assets or ecosystem functions and the productive base of the resource' [punctuation corrected from the original].
5. There is an interesting question about whether this national water policy requirement pertains to scientific 'information' or 'knowledge' or just to 'science'. The 1996 National Principles refer to 'scientific information'. The NWI refers variously to 'science' (para. 36) and 'scientific knowledge' (Schedule E, clause 6(iii). These terms may mean different things: see Section 2.2.
6. *Water Act 1989* (Vic.), section 4B, inserted by the *Water (Resources Management) Act 2005* (Vic.), section 4.
7. COAG (2004a, NWI para. 9) provides for the making of implementation plans. The accredited implementation plans are published on the National Water Commission website: http://www.nwc.gov.au/nwi/nwi_implementation.cfm.
8. Similar problems were foreseen in relation to the 'macro' planning process for the remaining water systems in NSW (National Water Commission 2006a: 2.27).
9. *Water Management Act 2000* (NSW) Parts 2, 3 & 4. Minister's plans are made under section 50, the sole section in Part 4.

10 This explanation is based on a personal communication from Paul Simpson of the Department of Natural Resources and Environment, NSW, on 24 October 2006, and from the facts described in *Murrumbidgee Horticulture Council v Minister for Land and Water (NSW).*
11 This was the policy decision of the NSW Cabinet in 1996.
12 It should also be noted that the equivalent of the general-security entitlements in Victoria are substantially more secure than in (Professor Kath Bowmer, pers. comm., 31 October 2006).
13 That trade constituted less than 1% of the total amount of water applied in the pilot project area, and was much less than the intrastate trade.
14 A discussion of the plans is available on the NSW Department of Natural Resources website: http://www.dnr.nsw.gov.au/water/sharing_plans0607.shtml (accessed 28 October 2006). The historical use has been calculated on the basis of the 'long-term average annual use' (Associate Professor Phil Eberbach, Charles Sturt University, Wagga Wagga, NSW, pers. comm. to Professor K Bowmer, October 2006).
15 See note 2.
16 This argument must be seen in a broader context of its legal feasibility: Jackson H (2005). Potential exposure to legal liabilities for the supply of recycled water and biosolids. *EPLJ* **22**: 418.
17 An example of a state program is on the NSW Department of Natural Resources website: 'water for the environment: water recovery projects', http://www.coag.gov.au/meetings/250604/index.htm (accessed 13 November 2006).
18 Australian Government, Productivity Commission, *Rural water use and the environment: the role of market mechanisms*, 11 August 2006, Box 8.3, p. 195, discusses the activities of the NSW Murray Wetlands Working Group in managing an adaptive environmental water entitlement.
19 The *Water Management Act 2000* (NSW), section 35(1), requires a plan to include 'performance indicators to measure the success of those strategies [i.e. the strategies to reach the objectives of the plan]'.

Chapter 5

Water planning

Principles, practices and evaluation[1]

Geraldine Gentle and Chris Olszak, URS

While intergovernmental agreements for the National Water Initiative (NWI) and the previous COAG water reforms have set the strategic direction for water policy in Australia, much of the detailed delivery is through water plans. Statutory water plans provide the operational framework for delivering much of the NWI and other water reforms. They set out catchment- or system-wide allocations for consumptive, environmental and other public purposes, providing the basis for individual water users' entitlements as well as the management arrangements for both surface and groundwater systems. Many of the challenges to the effective delivery of the NWI are reflected in water plans and planning processes. This chapter will discuss insights from recent water planning experience about how social science research and knowledge could improve water planning and thereby enhance delivery of the NWI.

Water planning is not new, but the water reform process has transformed the policy context and framework throughout Australia. Water planning was seen as a technical task in the past, the preserve of engineers and hydrologists, whose fundamental brief was to develop and manage surface and groundwater systems for beneficial use. Since the water reform process began, water planners have faced significant challenges in adapting and contributing to the new wider policy agenda. From looking at these challenges in water planning, it is clear that greater input from the social sciences – economics, politics, psychology, sociology and anthropology – could improve planning processes and outcomes around Australia.

Unfortunately, water planning processes and experiences are not well documented, at least in the public domain. Water plans are also very diverse, and the approaches taken by individual jurisdictions have been quite different. This chapter therefore relies on personal knowledge of the authors and water planners. Its generalisations may not be true for every plan or jurisdiction. Our objective is to identify areas where the social sciences can contribute to improving water planning in Australia as a whole, even if those needs are not so acute in certain individual cases.

1 Water planning in Australia

Statutory water plans sit at the centre of a broad framework of planning and regulation relating to land and natural resource management and use, although the systems differ in detail in each jurisdiction. They form not so much a hierarchy of plans as a complex network, with multiple linkages among plans and regulations of various forms and levels. The range of water plans includes:

- state/territory NWI implementation plans accredited by the National Water Commission (the Commission), including jurisdictions' commitments about water planning

- statutory water allocation and management plans. They cover a specified area, such as a regulated management area (e.g. the Lachlan regulated system) or a total catchment (e.g. the Fitzroy Basin) and are for a specified period (e.g. 10 years)
- water resource management plans, which may be statutory or not, depending on jurisdiction
- plans covering individual entitlements, separation of water from land and trading arrangements, such as Queensland's resource operations plans
- integrated natural resource management plans for a specific catchment (e.g. in SA)
- statutory regional land-use and development plans that require, link to or include water access, infrastructure and management arrangements for multiple uses, in a region such as Sydney or south-east Queensland
- government or intergovernmental basin (e.g. Murray-Darling Basin Commission's [MDBC] Living Murray) or state plans and strategies for water supply and management (e.g. Victoria's strategy for water for Melbourne and Central Victoria)
- non-statutory catchment management and integrated resource management plans with significant environmental objectives
- non-statutory plans by water supply authorities and companies for various activities, including management, infrastructure, delivery of water and waste water, flood management and recycling
- non-statutory plans and codes of practice by user industries and groups relating to their water use, sometimes in a co-regulatory framework
- individual users' water management plans, often a regulatory requirement.

Statutory water plans are the focus of this chapter, as they have a particular impact on delivery of the NWI. They are generally 'made' by the relevant state minister, sometimes with other ministers' input. They are a form of subordinate legislation.

These and the other types of water plans are linked with and through the overarching legislative and regulatory framework. Together they create a network of regulation and quasi-regulation that provides the basis for water allocation and management across Australia. These plans tend to have been developed at different times, for different purposes and under different legislation. While some jurisdictions have developed suites of consistent 'nested' plans, this is not the case in most other jurisdictions. This can create complexity, conflicts and anomalies in water and land use. This is a topic on its own and cannot be pursued here, but could be a useful topic for research.[2] Plans are only as good as their implementation and administration. Consequently, the institutional arrangements and resources available for water planning, implementation and review can also be significant issues, but are only touched on briefly here (see Chapters 8 and 10, this volume).

2 Water planning and water reform

Water planning is the pointy end of water reform in Australia, and as such, exposes many of the strengths, weaknesses and opportunities of the water reform process. Many of the commitments of the COAG Water Reforms and NWI intergovernmental agreement relate to and are reflected in water planning. The ability of water plans to implement these high-level principles largely determine the success of water reform in Australia. The Murray Darling Cap, environmental flows and exchange rates and estimates of losses for water trades all find expression in water plans of various types.

A key component of water reform in Australia in recent years has been the redefinition of water entitlements, allocation and trading regimes. Water plans of various kinds have provided the regulatory basis for the formalisation of access rights, the separation of titles to water from those for land and the basic rules for water trading in a catchment. Through statutory approaches such as the bulk entitlements process in Victoria, water-sharing plans in NSW, water allocation plans in SA and water resource plans in Queensland, access rights to water have been formalised and institutionalised. With evidence of significant ecological decline in many catchments, the process of strengthening entitlements (property rights) was combined with the objective of securing water for non-consumptive environmental and other public purposes. These dual purposes have added significant complexity to the planning process.

The water reform process sparked a rush of water planning around Australia. Every jurisdiction enacted new water legislation, and new water plans had to be developed reflecting the new principles. This has created enormous challenges for water planners. There have been significant gaps between the high-level principles of the NWI and the operational world of the water planner. The principles have had to be translated into operational guidelines for water planners. Often these guidelines were being developed in parallel with water plans or in response to the problems that the planners were experiencing.

The NWI is part of Australia's microeconomic reform program, but water planners typically are not familiar with that reform history or philosophy, nor with its language. They have had to learn a new language; the language of central government agencies. They have had to come to grips with the application of economics to natural resource management. They have had to interpret the principles of the NWI in a way that would be legally robust enough to stand up in court. Translating the high-level principles into working-level operational policies and guidelines for planners and stakeholders would not only help the water planning process, but would help to clarify the objectives of the agreement.

Water plans were important parts of the implementation schedules developed by each jurisdiction for the National Competition Council, and refreshed for the NWI and the National Water Commission. Previously, water planning had proceeded at different rates in the jurisdictions, but by the late 1990s everyone was at it. This was felt in a shortage of skilled and experienced technical staff. Suddenly engineers who had spent their lives designing or managing dams and irrigation systems became water planners charged with implementing major elements of the water reform agenda. These planners bravely embarked on the task, but had little information to guide them. While they had new legislation and the ministerial statements that went with it, they generally did not have a complete suite of operational policies for incorporation in water plans. This has meant policy development on the run and delays as policies have been developed, often with imperfect processes and results.

Nowhere has the disparate strategic view of the NWI and the water planner been more sharply focused than in the understanding of the role of economic instruments for water planning. Water planners have generally had little understanding of economics, and the senior officials who drafted the intergovernmental agreements tended to be economists with little understanding of water planning. So, despite the focus on economic instruments in the NWI, the understanding of economics brought to bear in water planning has been patchy at best. Little use has been made of economics as a policy instrument to achieve water planning objectives. Nor has there been much understanding of the implications for water planning of the use of economic policy instruments, such as pricing or trade. The use of economics has generally been limited to socio-economic impact assessment, usually done at the last minute as a compliance activity. Often these exercises were not embedded sufficiently within the planning process, which led to them becoming subjective or advocacy pieces driven by specific stakeholders, rather than objective, independent and integrated assessments.

Another tension between the NWI and water plans has arisen from the limited life of plans as entitlements have moved from a similar limited term with limited negotiability to perpetual tradable rights. This issue too will be caught up in the reviews of plans as they reach the end of their statutory lives. As each jurisdiction's legislative approach was different, no two planning processes were the same. While there was some exchange of ideas and information among the states, it was relatively limited except in the pilot water trading area of the lower River Murray – an initiative of the MDBC. Within some jurisdictions, the thousand flowers were also allowed to bloom, with different approaches adopted by each planning group. Timelines stretched out. Estimates of the time and resources required (and the commitments to the National Competition Council/Commission) turned out to have been optimistic. Some of the complaints about the slow pace of water reform relate to the slower-than-anticipated delivery of water plans.

In some cases, particularly in unregulated systems, scientific data were poor. However, hydrologists and engineers are used to working with less-than-perfect information, particularly in the jurisdictions most affected. The major sources of delay have not been technical, but have stemmed from:

- unrealistic timetables, often set at a high level of government without advice from planners themselves
- lack of comprehensive and realistic overarching policy frameworks, particularly relating to adjustment in overallocated systems (compensation, adjustment assistance, clawback mechanisms etc.)
- lack of practical and realistic operational policies in relation to economic, social and cultural issues
- inexperience of technical planners in dealing with complex economic and social issues and processes
- inexperience of technical planners in dealing with community engagement, particularly in situations of conflict and distrust
- inexperience of technical planners in dealing with policy processes and ministers
- inexperience of stakeholders and community groups in dealing with conflicts with government and water planners over science, economic and social impacts, values, information and institutions.

Not surprisingly, with planners learning by doing, they often found they had to develop operational policies on the run. However, the biggest issue that slowed water planning processes in fully or overallocated systems were about clawback of entitlements, adjustment assistance and compensation. The NWI optimistically states:

> water plans have been transparently developed to determine water allocation for the entitlements; regular reporting of progress with implementing plans is occurring; and a pathway for dealing with known over allocation and/or overuse has been agreed (NWI para. 47).

In one sense this is true. The de jure position of governments was simple: annual licences entailed no permanent property right, so there was no legal requirement to compensate irrigators for water clawed back to provide sustainable environmental flows or for other public purposes. However, the de facto position facing regional water planners was that it was considered economically, socially and politically unacceptable to clawback more than a marginal amount of water without some form of adjustment assistance. By and large, the states did not have programs in place to deal with the scale of adjustment required in severely overallocated systems, or in areas suffering significant adverse landscape change (e.g. salinity or waterlogging). Across the country, officials knew never to mention the 'C' word – 'compensation'.

This left water planners in an invidious position. They had no policy tools to deal with the fundamental issue. Technical water planners were usually not equipped to objectively analyse trade-offs among economic, social and environmental uses of water resources. Nor were they prepared for dealing with the highly charged economic, social and political issues, the negotiations and conflicts among stakeholders or the political processes. They have been water specialists, and have not been familiar with Australia's microeconomic reform experience in dealing with these issues in other sectors of the economy. Neither have they been helped by their community engagement strategies. Political scientists and sociologists have long known that local stakeholder groups themselves find it difficult to reach consensus on difficult decisions involving significant cuts in entitlements or allocations to their friends and neighbours.[3] In addition, most water planning frameworks have not critically reviewed the location of regional water-using industries and communities. Except for some small and isolated examples, such as in parts of northern Victoria, there has been little serious analysis of the most economically efficient spatial design of water use within states, let alone within river basins.

In the face of such intractable divisions, technical issues, such as the adequacy of the science, whether it is the hydrological modelling or the ecological condition and trend, become battlegrounds as well. Often, they are proxies for the real issues. Sometimes they are part of delay processes designed to protract or distract the planning process. In such circumstances, the temptation to develop temporary fixes is often overwhelming, allowing stakeholders to focus on areas of agreement and leave major reallocation issues for another day. However, this delays reallocation and improvement in environmental conditions. It puts additional pressure on the review process to address the overallocation problem. This will become more difficult as other aspects of water reform strengthen property rights and enhance tradability, shoring up the status quo.

While water plans are statutory plans and are made by ministers, water planners are not generally used to dealing with the hurly-burly of political processes. The allocation issues in stressed systems have inevitably entered the political arena, with interest groups lobbying ministers and MPs and attempting to alter draft plan provisions or planning processes. These issues could have been foreseen in overallocated systems and processes adopted to minimise adverse affects. However, participatory planning processes were often used in the hope that trade-offs could be reached on the ground. Consideration of the literature on participatory planning and the emerging fields of social science research encapsulated under the term 'political ecology' could have prevented this false start. However, the failure to adequately consider the political aspects of planning has caused deep divisions, suspicion and dissatisfaction with the eventual water plans that will have to be addressed when they are reviewed.

By contrast, when planning is being conducted with greater cohesion among the stakeholders, they tend to take the state of the science in their stride and make the best of it, agreeing to revisit issues as knowledge improves. Water plans have been completed relatively quickly in catchments where stakeholders had common rather than conflicting objectives. For example, the Cooper Basin Water Resource Plan was completed quickly in spite of limited scientific information, as graziers, environmentalists, Indigenous communities and government all agreed on fundamentals, particularly that irrigation should not be developed in the catchment.

None of these barriers and challenges should come as any surprise to those who have been involved in the experience of microeconomic reform in other sectors. The Productivity Commission's predecessors, the Industry and Industries Assistance Commission and the Tariff Board were all embroiled in debates about the adequacy of data and models as industry assistance was being dismantled. Reform to the dairy industry, fisheries and forestry industries have been through similar processes. All have involved complex economic, social and regional impacts and the development of adjustment assistance packages. Most have included significant adjustment relating to closure of industry in locations where it is no longer efficient. Much has been learned about policy measures to promote adjustment that could have informed adjustment policies, if they had been developed. However, this experience is not generally known to those involved in water planning, be they in government, user industries, the environmental movement or NGOs.

Water planning would benefit greatly from knowledge inputs from the social sciences that help to address the issues outlined above. This does not require fundamental strategic research. Rather, the needs are for more applied knowledge products, improving knowledge about planning processes themselves, participatory decision-making, capacity-building in the social sciences for technical water planners, and for adapting the lessons of microeconomic reform in other sectors to water allocation and management.

3 Limited life of plans and adaptive management

> A plan duration should be consistent with the level of knowledge and development of the particular water source (NWI, Schedule E).

States have limited the lifespan of water plans, typically to ten years, to enable adjustments to allocation among various uses without compensation. Limiting the life of water plans was supposed to give governments an opportunity to monitor and implement adaptive management. This is a pragmatic approach considering the uncertainty associated with climate variability/drought, possible impacts of human-induced climate change, community concerns regarding the long-term environmental sustainability of current levels of extraction, and the uncertainty associated with hydrologic and water use data underpinning the allocation regimes. However, as noted above, a number of plans in overallocated systems have left significant reallocation between consumptive use and environmental flows to the review process. An adaptive management framework should not be expected to deal with fundamental and major overallocation issues.

Limiting the lifespan of the water plans provides opportunity for reallocation of entitlements for environmental and other non-consumptive purposes if the allocation regime does not meet its objectives or is no longer considered appropriate. The NWI attempts to provide a consistent basis to address this challenge in relation to water plans through the risk assignment framework, which will be an important driver in the review and audit of water plans 'In the case of ongoing plans, there should be a review process that allows for changes to be made in light of improved knowledge' (NWI, Schedule E).

In NSW, for instance, the first set of statutory water-sharing plans which cover the majority (around 80%) of the state's surface water resources are due for their first review in 2014, a role assigned to the Natural Resources Commission under the *Water Management Act 2005* (NSW). Such reviews could engender vigorous debates among stakeholders, e.g. irrigators defending entitlements in the face of a number of years of low allocations and environmental groups wanting a greater share to address declining river health.

There is potential for debate over the level of proof and scientific certainty required to justify any change regardless of the risk assignment framework, even where it is incorporated in legislation, as in NSW. The review process is also likely to open up questions about the adequacy of the initial allocations, the planning processes used, the objectives of the plans and numerous other issues.

4 Water planning methodology

Modern water planning to implement Australia's water reform agenda is not the subject of textbooks and university courses. It is learned on the job separately in each jurisdiction. Much information about water planning issues is held within government agencies, but it is not readily or widely available. Jurisdictions have been experimenting and learning by doing largely in isolation from each other. While they may grapple with similar issues and use similar hydrological models, their policy frameworks, administrative processes and individual circumstances are all different, so that plans and their contents as well as the processes and resources for implementing them are also different. Similarly, the ways water plans link within the broader regional and natural resource management planning and regulatory frameworks vary across jurisdictions.

Jurisdictions have much to learn from each other about all aspects of water planning. It would be particularly useful to collate information from the various jurisdictions and to assess, compare and benchmark the water planning regulatory frameworks and processes in all the jurisdictions so that jurisdictions can learn from each other about:

- sharing modern water planning experiences and approaches
- standards and procedures for water planning
- setting reasonable timelines
- transparency in planning processes
- consistency and alignment of administrative processes used to implement plans, including property rights, compensation arrangements for enterprises or households adversely impacted by change, trading frameworks and land use planning
- stakeholder engagement – tools, best practice approaches, when to stop
- guidelines for the use of mediation, negotiation and other conflict resolution techniques
- developing the role of industry in water planning
- improving communication of the science behind water plans
- incorporating risk assessment and management in water planning
- implications of separate planning and management of surface and groundwater – managing the user arbitrage
- integrating different types of plans relating to water.

This could help to improve and align the regulatory design of water plans, the operational policy frameworks in which they sit and the administrative processes for implementing the plans. Similarly, it would provide useful information about the different approaches to planning processes, including:

- economic and social analysis for water plans
- structural adjustment and compensation for adverse impacts
- legal frameworks, property rights and issues
- communications with industry and communities
- stakeholder engagement, including negotiation and conflict resolution
- involvement of industry and NGOs
- relationships between planning and the political process.

This in turn would enable them to better assess the resources and time required for a particular planning process, as well as to improve the certainty and efficiency of planning processes.

5 Review and audit of policies and plans

End-of-plan reviews provide an opportunity to assess the appropriateness of the allocative balance between the environment and consumptive uses as well as the mechanisms used to allocate the resource. Balancing alternative uses, particularly in fully or overallocated systems, has been a matter of political judgment that would benefit from more rigorous analysis of the economic and social benefits and costs of alternative allocation regimes. The effectiveness of past performance against planned outcomes and purpose also needs to be assessed. Audits play an important role in distinguishing between the effectiveness of the plan itself, and how well it has been implemented. Both of these functions would be enhanced by greater use of the social sciences.

Clear and transparent objectives and a coherent monitoring and evaluation framework provide the basis for robust adaptive management decision-making. The literature on investment planning, program evaluation and adaptive management provides much useful knowledge relating to evaluation frameworks and guidance in application (such as the logical framework approach which links inputs to outputs and outcomes) (Commonwealth of Australia 2000). However, application of the principles has been patchy in practice and the concepts and language of evaluation are often used inconsistently. A key challenge in reviewing water plans is the difficulty of retrospectively defining objectives and the linkages between rules and expected outcomes. Uncertainty is unavoidable in such a complex arena, but making assumptions explicit and testing them through targeted research is a key ingredient that has been applied to a limited extent to date.

In this respect, perhaps there are some important questions from a social science perspective.

- What are the social, environmental and economic outcomes and purposes resulting from implementation of a water allocation plan?
- How would we know that these outcomes and purposes have been achieved?
- Who should measure the changes at outcome and purpose level?
- How should performance be communicated?
- How should the results be used in adaptive management decision-making?

These questions are inextricably linked. Defining broad social and economic objectives for water plans such as those linked in some way to the 'economic value of production' or the even more intangible social or community objectives relating to long-term 'viability' or 'sustainability' creates a number of attribution problems – even the best entitlement regime cannot dictate when droughts occur, or changes in underlying industry profitability. It can be difficult to define an appropriate base for comparison in the face of natural variability. For example, ecological relationships are known to display complex responses, lengthy temporal lags and thresholds. While continuation of past practices may have led to more rapid decline in ecological condition, it can be difficult to demonstrate improvements due to the plan over a relatively short period. From a social and economic perspective, failure to place a limit on extraction may have led to an unsustainable access regime with significant overinvestment in an industry and insufficient water to meet requirements. This could have resulted in significant adverse economic and social impacts to both new and existing resource users.

This point emphasises the potential benefit of translating the high-level objectives of water reform into more operational objectives that are more closely aligned with the direct purpose of allocation plans, such as specific objectives related to access, security and transferability of entitlements. Research is required to develop cost-effective techniques to measure the effectiveness of these more focused objectives through, for instance, attitudinal and behavioural surveys with entitlement holders, as well as other indicators that reveal the presence/absence of such outcomes, for instance the level of investment. Equally, social science research may help us understand how to assess the community's level of confidence in the security of the environment's entitlement. A critical precursor to attempting to measure perceptions of security in the entitlement will be determining the extent of understanding the basis, role and function of the entitlement, its duration and conditions.

Attitudinal research and applied case studies may also have a role in demonstrating the links between improved security of entitlement and investment on-farm or trade to higher-value uses. However, these are leading indicators that cannot demonstrate effectiveness – only measurement of changes at outcome and purpose level can do that, and these are necessarily lagging indicators. Thus the adaptive management cycle for water resources is prolonged. There may be other indirect research that could provide information about perceptions of security of entitlement. For example, understanding attitudes driving the low levels of trade in permanent entitlements could be a function of remaining uncertainty in the water reform agenda. Chapter 6 in this volume deals with this as one of the impediments to water trade.

The underlying philosophy behind the current 'entitlements' approach to water allocation is that it will naturally produce sustainable and efficient outcomes through the creation of a valuable asset. The term 'entitlement' in itself creates a semantic issue. What was adopted as a means of distinguishing individual water users' shares from the overall allocation, can create the wrong mind-set. In the longer term, it may be necessary to revert to broader outcome indicators to assess the effective use of water resources including ex post examination of changes in the economic value of water, assuming that a market of some sort exists. However, measuring the economic value of agricultural production ex post is also data-intensive and subject to a number of methodological challenges, including the appropriateness of continuing to rely on gross margin estimates, and it defies trends towards reducing the frequency and type of agricultural data collected.

This example demonstrates a more general issue in the development of a robust monitoring and evaluation framework: what data should be collected and analysed? Governments would benefit from research which examines the benefits of collecting each dataset with the cost of obtaining it. Research and development of cost-effective methods of monitoring and assessment using GIS and other new technologies could also contribute positively to the review process (see Chapter 8 on Integrated Assessment).

5.1 Performance audits and ongoing monitoring: engaging the community and sending the appropriate signals

Performance audits of water plans are generally considered to focus on compliance and implementation of the plan (measuring inputs, activities and perhaps outputs – none of which provide data for analysis of effectiveness). In NSW, implementation plans are being developed by the Department of Natural Resources for the initial batch of 36 water-sharing plans, with audit plans to follow. A key challenge for policy-makers in coming years will be to further consider how high-level objectives and shares in the resource defined in plans actually translate into operational instructions, and then how these are actually put into practice. Significant improvement in water allocation management could arise simply from ensuring that the plans are actually put into practice as intended. Performance and implementation audits could play a significant role in meeting this need. Issues of cultural change in agencies responsible for operational water management provide an interesting area of social science research.

The audit of a water plan's performance and implementation requires significant expertise and knowledge of operational water resource management. Areas of possible research interest to social scientists include the methods of ongoing community engagement in monitoring, evaluation and reporting of water plans. Research could examine:

- cost-sharing for water plan implementation
- the role of the community in the process of implementation audit and review
- the most efficient way to conduct and benchmark performance audits to create a competitive environment for investment in water plan implementation
- the role of the community in ongoing monitoring and reporting – whether community involvement in environmental monitoring could build knowledge, capacity and understanding of allocation decisions
- the impact of audit and monitoring results on perceptions of entitlement security and flow-on impacts to investment – keeping industry informed could signal possible outcomes associated with the end-of-plan reviews.

5.2 Using review results to make decisions

Ultimately, at the end of each water plan period, governments are required to decide whether changes to the current entitlements are appropriate, decide who will bear the cost of any reallocation and how to allocate any new investment resources between competing water plan proposals. Even if a water plan has clear and measurable outcomes and purpose indicators and good monitoring data and analysis demonstrating its effectiveness, governments will ultimately need to decide whether the balance between the environment and consumptive uses agreed to ten years prior is still appropriate. Typically, ministers will want to be certain that any adverse impacts will be outweighed by the positive impacts.

Thus the decision-making process, such as that of turning a review finding associated with effectiveness into a decision about appropriateness and any associated reallocation, is critical. Transparency and community engagement in the process will play a key role in determining the acceptability of decisions. This process should be designed, communicated and agreed before the reviews commence, with considerable effort focused on designing the preconditions and criteria that will guide change – i.e. 'if desired outcome 1 is less than X, then we will do Y'. Applied research in this area is vital.

5.3 Addressing the question of appropriateness through ex ante scenario modelling

Whether 'the balance is right' is extremely difficult to answer objectively if we rely on ex post review findings. An ex ante consideration of the likely future outcomes by modelling various allocation scenarios, including continuation of the current regime, may be an effective complementary area for water planning. This could be achieved through an integrated assessment framework (as described in Chapter 8) or what natural resource economists would call integrated benefit cost analysis – where environmental, social and financial impacts are considered from the perspective of social welfare. From a functional perspective, this process needs to be separated from the review of past performance and there is an important question regarding whether it should be considered part of the development of a new plan.

A predictive modelling approach would require specification of relationships between hydrological outputs and environmental, social and economic outcomes. Applied research in this area is required as there is significant complexity in these relationships. From an economic perspective, further applied research is required to understand the marginal value of water in alternative uses including agriculture, mining, urban supply, non-use environmental/ conservation values and recreational values (Young 2005). The integrated benefit cost analysis approach was used in Victoria to assist in revising bulk entitlements (particularly for the provision of greater environmental flows) in the Thomson and Macalister (Branson et al. 2005) as well as the Tarago River systems. In both catchments, environmental benefits were compared to economic costs to both irrigated agriculture and urban supply security. Cost-sharing implications were also considered. Improved methods for considering the regional importance of water-intensive industries and the possible impacts on community viability could improve the link between direct economic and broader social impacts.

6 The research agenda

The greatest challenge in dealing with the issues identified above is not the need for new fundamental research. Many of the issues identified could be rectified by the development of improved information, communication and training products and programs. Incorporating economic and other social science analysis into water planning frameworks and processes at inception can tackle many of the other issues identified. This can be done by both developing more multi-disciplinary planning teams, and by increasing technical planners' appreciation of the potential role of the social sciences beyond socio-economic impact assessment. It is our assessment that there are six key areas where the social sciences can contribute.

The first need relates to the documentation and assessment of water planning. Basic documentation of water planning requirements, methodologies and processes in Australian jurisdictions is required, at a minimum, and there is a significant need to identify and develop the standards and procedures for water planning in each jurisdiction. Related to this is the adequacy of operational policies and planning processes, and the cost-effectiveness, timeliness, transparency and level of stakeholder engagement in those planning processes.

The second need concerns the improvement of water planning in each jurisdiction. A benchmarking product for evaluating the efficiency and effectiveness of planning and the implementation of water plans is needed, which could include a series of practical guides (and related case studies) to the NWI for operational staff and stakeholder use. These guides might include basic information concerning the objective and key elements of the NWI, and more detailed information on designing and implementing planning and review processes including appropriate and effective participation; a checklist of operational policies used in jurisdictions for incorporation in water plans, so that jurisdictions can identify and fill any gaps before starting planning processes; information on the use of economic instruments as instruments of water planning and on water plans as regulatory instruments; and information on the legal issues in implementing the NWI, including the risk assignment framework. Such an initiative would greatly increase water planners' capacity to integrate better water plans with regulation and the overall planning framework, and to gain a better understanding of economic and social impact assessment in water planning.

The third area pertains to the need for capacity-building among water planners, which should include training programs for operational planners. In addition to the material identified above, there is a need for a practical guide to analysing trade-offs in water allocation and management (i.e. benefit–cost analysis) and perhaps a national water planning workshop aimed at middle-level management to share experiences and suggest improvements in planning arrangements. Similarly, a guidance manual on performance auditing, monitoring, supervision and evaluation of water plans and their implementation incorporating uncertainty would prove useful.

The fourth area relates to the review process for water plans. The water planning framework has been developed to allow for 'adaptive management' changes to water allocation and management arrangements as a result of improvements in knowledge or changes in resource condition. In pragmatic terms, adaptive management provides some leeway to allow authorities to make changes they consider necessary or acceptable. However, in principle, adaptive management should be a well-defined response to implement clearly specified objectives. This research area would examine the concept of adaptive management as applied to water planning, with a view to developing understanding of the concept and the ways it could be applied more rigorously. Underpinning this is the need for national water resource management planning and implementation benchmarking frameworks for the review of water plans and their implementation. Similarly, a framework for performance audits and benchmarking the administrative and economic efficiency of water resource management planning and implementation should be developed.

The fifth area of research relates to the need to better integrate water planning with the objectives of the NWI. Ongoing periodic reviews of the NWI's approach to water planning, in light of experience, would be useful and could provide the basis for evaluation frameworks for reviews of water plans such as the 2007 review of state/territory implementation plans. In line with the recommendations discussed in Chapter 8, these evaluations would include the scope, process, methods and data needs for better decision-making, which would in turn contribute to the development of benchmarking products for water planners.

Finally, there is a very strong need to better understand the relationships between water planning and reviews, decision-making and uncertainty. The water planning, audit and review processes are subject to considerable scientific uncertainty, given the volatility of the Australian climate, the long duration of landscape processes and gaps in scientific knowledge about surface and groundwater systems and ecological conditions. The use of inadequate information in planning leads to a lack of confidence in recovery programs or unwillingness by stakeholders to sacrifice production for uncertain gains. Climate change is exacerbating this. Water planners would be assisted by manuals and training programs on incorporating uncertainty into water planning, audit and review. Similarly, the precautionary principle is not well understood, but when conflict arises because of competition for limited water resources for consumptive and environmental purposes, it is even more difficult to prevent its misuse and distortion. It could be useful to explain the principle in terms of water planning issues.

References

Boully L, McCollum B, Vanderbyl T & Claydon G (2006). Talk until the talking starts: resolving conflict through dialogue. Manuscript.

Branson J, Sturgess N & Dumsday R (2005). Economics of environmental flows: benefit cost analysis of environmental flow options: Thomson and Macalister Rivers. *Water*, September, 35–39.

Bromley DW (1992). *Making the commons work. Theory, practice and policy.* Institute for Contemporary Studies: San Francisco.

Claydon, Greg (2006). Personal communication.

Commonwealth of Australia (2000). *AusGuidelines: the logical framework approach.* AusAID: Canberra.

National Water Commission (2006). *2005 national competition policy water reform assessment.* Available at www.nwc.gov.au/nwi/ncp_water_reform.cfm.

New South Wales Department of Natural Resources. *Water sharing plans.* Available at www.dnr.nsw.gov.au/water/plans.

North DC (1990). *Institutions, institutional change and economic performance.* Cambridge University Press: Cambridge.

Ostrom E (1990). *Governing the commons. The evolution of institutions for collective action.* Cambridge University Press: Cambridge.

Peterson D (2006). Precaution: Principles and practice in Australian environmental and natural resource management. Presidential address presented at the 50th Annual Australian

Agricultural and Resource Economics Society Conference, Manly, New South Wales, 8–10 February 2006.
Queensland Department of Natural Resources and Water (2006). *Water resource plans.* Available at www.nrw.qld.gov.au/wrp.
Robbins P (2004). *Political ecology*, Blackwell Publishing: Oxford.
South Australia Department of Land, Water and Biodiversity (2006). *Water allocation plans.* Available at www.dwlbc.sa.gov.au/licensing/wap/index.html.
Tasmania Department of Primary Industries and Water (2005). *Generic principles for water management planning.* Water Resources Policy # 2005/1. Available at www.dpiw.tas.gov.au/inter.nsf/Attachments/JMUY–67X5ZY/$FILE/Generic%20principles%20WMP.pdf.
Victoria Department of Sustainability and Environment. *Victoria's water allocation framework.* Available at www.dse.vic.gov.au/DSE/wcmn202.nsf.
Western Australia Department of Water (2006). *Draft state water plan.* Available at http://portal.water.wa.gov.au/portal/page/portal/PlanningWaterFuture/StateWaterPlan/Content/Draft%20State%20Water%20Plan.
Young RA (2005). *Determining the economic value of water: concepts and methods.* Resources for the Future: Washington DC.

Endnotes

1 This chapter combines two of the topics identified in Water Perspectives: first, 'Water plans: processes, content and accreditation' and second 'Audit and review of policy and plans: measuring effectiveness'. For each topic, the task involves exploration of the social science and economic research needs to progress the NWI. In exploring the second topic, the chapter continues to focus on water planning, specifically water allocation plans rather than 'water policy' in its broader sense. The research needs associated with the review and audit of these plans from a social and economic perspective are directly transferable to other plans and policy more generally.

2 For instance, further research in Australia could build on the extensive international literature on governing natural resources (Ostrom 1990), institutional economics (Bromley 1992; North 1990) and the broader field of political ecology (Robbins 2004) to improve our understanding of a multitude of concepts relevant to water planning, including 'nesting' and 'polycentric institutions'.

3 A fascinating account of the journey of water planners and users through conflict to understanding and achievement of a plan is given in Boully et al. (2006) in relation to the Condamine Balonne and Border Rivers water resource planning processes.

Chapter 6

Water trading and pricing

R Quentin Grafton and Deborah Peterson[1]

Water in most of Australia is an increasingly scarce resource. A growing population and climate change (Young et al. 2006), coupled with an increasing awareness of environmental costs of water withdrawals, will place an increasing burden on balancing supply and demand for water. Australian governments are acutely aware of the water management challenges. In 1994, the Council of Australian Governments (COAG) promoted innovations that included better-specified water rights, market-based approaches to water management, and the need for environmental flows, especially in the Murray-Darling Basin. These reforms, including a cap on withdrawals in the Murray River systems, were designed to ensure water is allocated to its highest-value use while ensuring ecologically sustainable development (Quiggin 2001).

The 2004 National Water Initiative (NWI) expanded the market-based agenda with its endorsement of better-specified water access entitlements that can be recorded in reliable and nationally consistent water registers (COAG 2004). The NWI also included agreement on risk-sharing of reductions in water allocations, but its implementation remains a challenge. The NWI endorsed development of water plans that are intended to ensure environmentally sustainable levels of extraction and, eventually, perpetual water access entitlements denoted as a share of a specified water resource. To ensure water is allocated to its highest value, parties to the NWI agreed to minimise transactions costs on water trades, promote efficient water markets and remove barriers to water trades.

The NWI will help address critical issues of overallocation, inefficiencies, inconsistencies and lack of coordination across jurisdictions, and inadequate water planning. It includes key milestones such as the immediate removal of barriers to temporary trades in water, implementation of compatible and publicly accessible and reliable water registers by the end of 2006, and the establishment by 2007 of compatible institutional and regulatory arrangements to facilitate intra and interstate water trading.

In this chapter, we provide our understanding of the current state of play in terms of water trading and pricing, what can be learned to improve the reform process, and what gaps must be addressed. Section 1 reviews the state of water markets in Australia, paying particular attention to constraints on trade and the benefits of increased trading. Section 2 examines key issues in terms of water prices, including pricing environmental externalities and the use of scarcity pricing in urban centres. In section 3, we use experiences from fisheries to offer insights for the water reform agenda; section 4 briefly reviews key knowledge gaps.

1 Water trading

Water trade in Australia is largely limited to transactions between irrigators. The most active trading region is in the southern Murray-Darling Basin, but markets have expanded considerably since the mid 1990s and should develop further with NWI initiatives.[2] Surface water trading by irrigators, particularly in seasonal allocations, is widespread in many irrigation districts. Markets for derivative products for water, such as leases and forward contracts, are emerging, largely in response to irrigators' preferences for more flexible trading arrangements.[3]

Initially, water trading was restricted to trade between irrigators within the same irrigation district. Over time, trade expanded to include intravalley and interstate trade. Interstate trade, however, is restricted to regions in the Pilot Interstate Water Trading Project. Figure 6.1 shows intrstate and interstate trade in seasonal allocations, and Figure 6.2 shows intrastate and interstate trade in water entitlements, both in the southern Murray-Darling Basin. Interstate trade in water entitlements is likely to expand: at the recent summit on the southern Murray-Darling Basin, the federal government and the governments of Queensland, SA, Victoria and NSW agreed to ensure that permanent interstate trading will commence in the southern Murray-Darling Basin states by 1 January 2007.

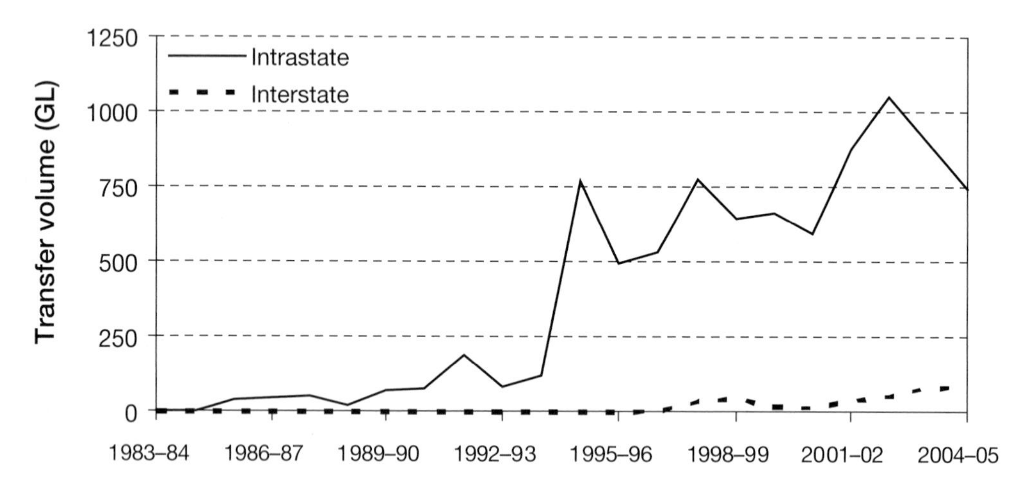

Figure 6.1 Seasonal allocation trade in the southern Murray-Darling Basin.
Data source: C. Biesaga, MDBC, pers. comm. 26 May 2006, cited in Productivity Commission (2006).

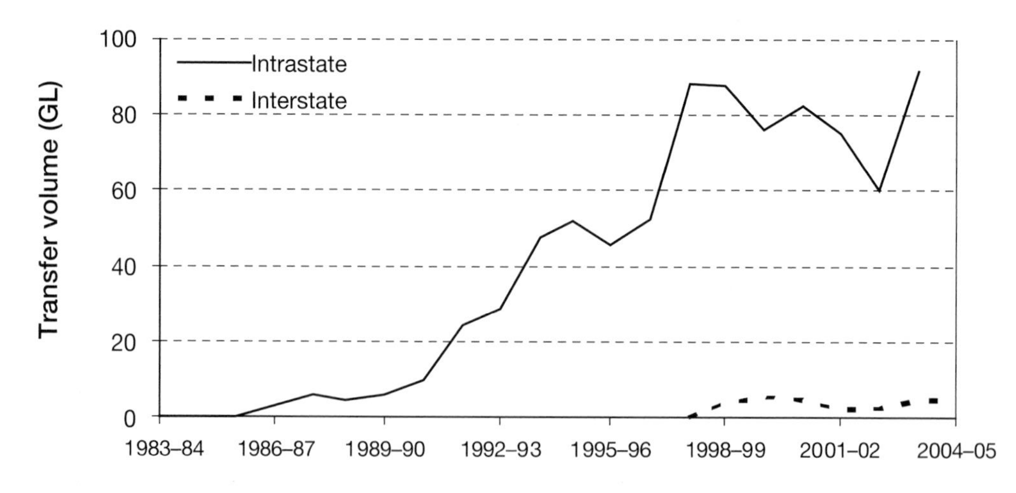

Figure 6.2 Water entitlement trade in the southern Murray-Darling Basin.
Data source: MDBC, pers. comm. 26 May 2006, cited in Productivity Commission (2006).

Publicly available data on prices and volumes of water trade over time and across regions are limited. The price of water entitlements varies temporally, with the attached level of security, and across regions. The Australian Bureau of Statistics (ABS) Water Survey of Agriculture 2002–03 provides a snapshot of traded prices. For temporary purchases, prices ranged from an average of $8 per ML (paid by some irrigated pasture farms in north-east Australia) to $1508 per ML (paid by some irrigated fruit growers in the interior of the country where there is low rainfall). Permanent trades in the same period were reported as ranging from $80 to $4819 per ML (ABS and Productivity Comission (PC) 2006). By contrast, water cost charges paid by irrigators to water authorities and supply companies range from $30 per ML to close to $90 per ML in the Murray-Darling Basin (Qureshi et al. 2005).

In the three main irrigation districts in the southern Murray-Darling Basin (Murray Irrigation, Goulburn-Murray Water and Murrimbidgee Irrigation), aggregate trade in seasonal allocations represented, on average, 11–20% of total annual allocations in 2000–01 to 2002–03. By contrast, trade in entitlements represents a much smaller share of allocations – on average, 1–2% across the three major districts between 2000–01 and 2002–03 (Peterson et al. 2004).

The phenomenon of a higher proportion of temporary asset transactions relative to permanent entitlement sales is not unique to water markets. Thus, by itself, it does not suggest market failure (PriceWaterhouseCoopers 2006). The number of participants in the market and the volume of water traded are determined by the supply and demand for water and by any restrictions on participation in trade. A small number of buyers and sellers may reflect a number of factors, including adequate existing water allocations relative to demand; homogeneous production, such that water demands are similar across irrigation activities; the inability to transfer water cost-effectively between potential buyers and sellers; and restrictions on water trade, including for environmental purposes (PC 2006).

The fact that sales of water entitlements are fewer than sales of seasonal allocations also reflects the inherent market uncertainties and risk associated with 'permanent' trade. In the absence of risk, the value of an entitlement is the sum of expected net returns from all future allocations. However, a permanent entitlement – unlike a seasonal allocation – provides a hedge against the risk of any future reductions in resource availability, and is a risk management tool for irrigators, especially those with large and durable fixed assets (Hafi et al. 2005).

1.1 Constraints to water trade

A recent report by the Productivity Commission (PC 2006) reviewed potential constraints on water trade, including non-regulatory and regulatory measures and administrative arrangements. Trade in seasonal allocations is relatively free in most large irrigation districts; many remaining restrictions reflect hydrology. Nevertheless, there remain some constraints for which the rationale is not always transparent. A wider range of restrictions applies to trade in water entitlements, with restrictions greater between districts than within. Non-regulatory constraints include hydrological factors, high transaction costs, limited market information and social factors. Not all of these constraints give rise to economic inefficiencies. In some cases, it is not currently feasible to remove non-regulatory constraints. For example, water trade requires connected (natural or built) infrastructure to facilitate the movement of water from the seller to the buyer, but as many catchments in Queensland are not hydrologically connected trade is restricted to schemes within a catchment.

Transaction costs are the costs of establishing, monitoring and enforcing property rights. All else equal, transaction costs tend to reduce trade as they reduce the net gains from selling and buying water (Connell et al. 2005), and can be substantial (McCann & Easter 2004). Transaction costs may be fixed in the sense that the amount paid is independent of the amount of water traded, or they may change with the amount traded (such as a commission paid to a water broker). They may be imposed by regulations or arise privately. For instance, the time and effort required by a willing buyer to locate and bargain with a willing seller represent transaction costs. Compatible and reliable water registers, as proposed by the NWI, should help reduce these costs.

Other relevant transaction costs include government fees and charges, approval activities, brokerage fees and utility charges. Transaction costs for trades in permanent entitlements are generally higher than those for seasonal allocations (se PC 2006). Most significant are the recently imposed exit fees on the export of water entitlements out of some irrigation districts, which are as high as 74% of the value of entitlements and impose a significant barrier to trade (Crase et al. 2000; PC 2006). Costs also vary significantly across jurisdictions. The Allen Consulting Group (2006) found that for trade in seasonal allocations, transaction costs as a percentage of the value of trade were 2.5–3.1% in NSW, Victoria and Queensland, but 21% in SA.

Limited information and lack of transparency of water trade prices and trading rules can impede water market participation. Trading facilities, such as those managed by Waterfind, Watermove and SunWater, are beginning to emerge and provide relevant information to traders. Social factors can also influence trade. Some regional communities have expressed concern about the consequences of exporting water out of a region, and there are reports of considerable social pressure being placed on potential water sellers not to trade (Fenton 2006).

Regulatory and administrative arrangements can constrain water trade within and between irrigation districts, as well as between types of users. Such constraints exist at both state and district levels, and are imposed by state or territory governments or by water utilities (including irrigation companies or cooperatives). Constraints include trading rules and zones, measures imposed to address stranded assets and other costs associated with structural adjustment, inefficient institutional arrangements, and excessive charges or slow approval processes. In some jurisdictions, legislation prohibits the purchasing of water by government agencies and non-landholders. In Victoria, for example, ownership of water is currently restricted to people owning or occupying land with access to individual water entitlements, and government agencies (including state-owned water utilities). From 1 July 2007, non-water users will be able to purchase water, but this will be limited to 10% of the maximum volume of entitlement in the particular water system. Such restrictions, along with policies discouraging certain organisations from participating in water trade, can represent substantial barriers to trade.

An often overlooked but nevertheless significant constraint on trade is the use of administrative arrangements within water plans to allocate water. Water planning frameworks allocate portions of the water resource pool to different uses, such as the environment, agriculture and urban activities. These arrangements also allocate water use within various subsets of the water resource pool, such as groundwater and surface water. There are opportunities to use markets to substitute for these arrangements to more efficiently allocate water among competing uses, such as through the encouragement of environmental water donations, auctions for ecosystem services and other approaches (PC 2006).

Apart from regulatory and administrative arrangements directly restricting trade, other arrangements governing water entitlements and allocations – such as jurisdictional progress in separating water entitlements from land, lack of integration in water management systems, poor water accounting systems and the number and complexity of entitlement types – also reduce water trade.

1.2 Effects of freeing up trade

The efficiency impacts of constraints on trade differ depending on their nature, extent and location. They will be greater over time, as resources fixed in the short term can be adjusted and moved to alternative uses. In general, the adverse economic impacts of trade restrictions are likely to be greatest where there are large differences in the water needs and valuations of water across water users and regions, and where water is relatively scarce and alternatives are not readily available. In other words, the constraints on water trade have the greatest impact where the potential benefits from trade are greatest. Consequently, restrictions or prohibition on water trading in droughts where water is available for use has high economic costs.

Heaney et al. (2004) and Peterson et al. (2004, 2005) examined the effects of removing constraints on water trading by irrigators in the southern Murray-Darling Basin. Both studies suggest that additional trade among irrigators may be relatively small, even if impediments are removed. However, the importance of trade is likely to increase over time as producers make new investments in higher-valued activities, and if more water is sourced from consumptive uses to meet environmental objectives (Heaney et al. 2004). Further, as noted by Peterson et al. (2004), the benefits of trade will be greatest in years of low water availability. This is because water reductions would have a larger effect on gross regional product (and water trade would provide larger gains through mitigating the impact on gross regional product) in years with low water availability.

Gains from removing some constraints on water trade

Heaney et al. (2004) assessed the economic impacts of water trade among irrigators under freer administrative arrangements and alternative charging options for rural water delivery. Using a partial equilibrium model of water markets in the southern Murray-Darling Basin, the study suggested that removing administrative impediments to water trade will result in around 600 GL (around 15% of total water entitlements in the region at that time) of additional trade in permanent water entitlements. Heaney et al. considered that this partly reflects the large sunk investment in on- and off-farm infrastructure. They suggested that the volume and value of entitlement trade will increase as infrastructure assets near the end of their life and as other water demands (e.g. demand for water for environmental purposes) increase.

Peterson et al. (2004, 2005) examined the effects of hypothetical reductions in water availability in the southern Murray-Darling Basin, under scenarios of no trade, intraregional trade (by irrigators) only, and both intra and interregional trade by irrigators. The study indicated that allowing both intra and interregional trade by irrigators would halve reductions in gross regional product due to a hypothetical decrease in irrigation water availability of 10% (from 1% – $356 million in 2003 – to 0.5%). The modelling approaches of both studies involved a number of simplifications and assumptions. Neither study considered the effect of changes in water trade on environmental conditions (such as salinity), nor the effects of environmental externalities or other third–party impacts on water trade.

Sources: Heaney et al. (2004); Heaney et al. (2006); Peterson et al. (2004, 2005).

Dwyer et al. (2005) used a general equilibrium model (TERM–Water) to examine the effects of expanding trade of water in south-east Australia to include both irrigators and urban users. The study demonstrated the reduced losses in gross regional product that can arise from a reduction in water availability when water trade is allowed. When regions with relatively low levels of water consumption, such as Adelaide and Canberra, face shortfalls in water and trade with regions that use large volumes of water (such as irrigators in the southern Murray-Darling Basin), they have little effect on traded prices and quantities of water. However, the study showed that if large urban water users, such as Melbourne, traded with a relatively small rural area such as Gippsland, the price effects could be substantial.

Young et al. (2006) also used TERM–Water to investigate the effects of allowing urban–rural trade. That study incorporated projections of population increases to 2032, and considered scenarios involving new sources of water, for example desalinisation. They found that allowing urban water utilities to purchase water from irrigators significantly reduces the potential increase in price that equates supply with demand. Modelling developments beyond TERM–Water are, however, required to distinguish between trade in seasonal allocations and trade in entitlements, and how choices may change within an irrigation season.

2 Water pricing

Competitive markets allow water to be traded so that those users with the highest marginal value (after accounting for transport and transaction costs) are able to purchase water from lower-value uses. This exchange is welfare-enhancing to the buyer and seller and promotes efficiency. Trading establishes the market price – that changes with environmental conditions and economic circumstances – which signals to all water users its relative value. However, if markets do not operate effectively then the price signal and the incentive it provides to water users is distorted. This reduces efficiency and can have important distributional effects. For example, if a large purchaser of water has market power in the sense that its actions affect the price paid, it is possible that sellers of water may receive less for their water than if the market were competitive. This benefits the buyer at the expense of sellers, can reduce the amount of water traded and may distort the price of other goods that use water as a factor of production. The high volume of trade in seasonal allocations in the southern Murray-Darling Basin suggests little cause for concern about the exercise of market power, but the lower volume of trade in permanent entitlements could allow opportunistic behaviour. Greater information about what water is available, such as through water registers, will assist in promoting competitive markets, as will greater transparency regarding the prices paid for water in different catchments and jurisdictions.

2.1 Externalities

Many different environmental changes are associated with the supply and use of water including changes in habitat, water quality and ecological conditions. These environmental costs are called externalities if the costs (increased salinity, reduced water supply downstream etc.) imposed on others (downstream users, recreational users etc.) are not accounted for by those withdrawing water for use. Where negative externalities exist, too much water is used relative to the benefits delivered and inappropriate signals are provided to water users in terms of their land and water-use practices. Under the NWI, COAG agreed to continue to manage environmental externalities through a range of measures. As well as adopting regulatory approaches, it was agreed to examine the feasibility of using market-based instruments such as pricing to account for both positive and negative externalities associated with water use, and to implement pricing that includes environmental externalities where feasible and practical (clause 65(ii)). At present, only SA and the ACT have explicit environmental management charges levied on water consumed (Dwyer et al. 2006), but in neither case is this directly related to the external costs imposed by water users.

However, environmental charges may not be suitable for all water-related externalities (Hatton MacDonald 2004). Dwyer et al. (2006) consider the complexities of implementing a tax or charge related to the costs of negative externalities. The externality's characteristics affect the likelihood of private sector solutions being effective, and influence the effectiveness of alternative public policy instruments. While a tax equal to the marginal external costs at each level of output could improve efficiency, and in the long term provide incentives to undertake abatement activities, the design of such a tax is not trivial.

2.2 Scarcity pricing

An important initiative of the NWI is urban water reform that includes the goal of efficient water use, improved pricing and a review of the effectiveness of temporary water restrictions (clauses 90, 91). Many urban centres offer the possibility of rural–urban water trading given the large disparity in costs to users, even accounting for the extra cost of water treatment and delivery to urban users (Quiggin 2006; Chapter 7, this volume). In addition to facilitating urban–rural trades, we need to reform urban water pricing to account for the large temporal variations in supply that are currently managed with quantitative water restrictions.

A recent study of Sydney water suggests that should a low rainfall period similar to that of 2001–05 occur again the city would become critically short of water, even with expected increases in supply from groundwater and recycling. An alternative to ongoing water restrictions is to have the price of volumetric water adjust flexibly upwards as the amount of water in storage declines (Grafton & Kompas 2007). Pricing urban water to balance demand with available supply is economically efficient, and signals users as to the value of water they are using (Sibley 2006). Such pricing also promotes innovation in terms of supply. The key point is that the price of water must reflect its opportunity cost in use and non-use, balance supply with demand and provide appropriate signals for conservation and efficiency.

3 Adaptable knowledge

The NWI places Australia in the forefront of worldwide policy reform in terms of water markets, planning and institutional developments. Although there is a great deal of knowledge about forwarding the place of water rights on the reform agenda (PC 2003, 2006; ACIL Tasman 2004; PriceWaterhouseCoopers 2006), much can be learned from other resource sectors that have initiated market-based reforms earlier.

The fisheries harvesting sector offers a number of possible insights for the water reform process. Fish are mobile and are not homogenous, and differ in size and type depending on where and when they are caught. The mobility and the spatial and heterogeneous qualities of fish have significant parallels with water. Moreover, as with water use, the capture of the resource frequently imposes cost on others, and how it is appropriated also affects the environment. The fishing industry in Australia and other jurisdictions has experimented with and refined the use of market-based instruments. These rights are typically defined as a share of the total allowable catch (TAC) and are called individual transferable quotas (ITQs). The use of these rights offers insights to the issues of overallocation, incentives and third-party effects in water reform.

3.1 Overallocation

The fishing analogy for surface water overallocation is overcapacity. In the case of fisheries, overcapacity refers to the case where the inputs used in fishing (vessels, nets etc.) exceeds what is required to catch the available quantity of fish. In the case of water, overallocation means that water infrastructure and existing water-use activities exceed what is feasible given the available water supply. Overallocation has environmental consequences because it makes it much more difficult to reallocate water for environmental flows when there is no 'surplus' water, especially in drought conditions. Similarly, in terms of overcapacity, fishery managers find it difficult to reduce fishing inputs to meet sustainability goals if this imposes hardship on marginal fishers. This process is much more difficult where uncertainties over the impacts of reduced flows collide with social and economic interests, and when ecological risk boundaries are neither clearly defined nor widely accepted (Grafton et al. 2007).

The way to resolve the overcapacity problem in fisheries has been to change the incentives faced by harvesters. By implementing ITQs, more-profitable fishers can acquire a greater share of the harvesting rights, thereby encouraging the exit of less-profitable operators. To be successful, however, the total harvest must be appropriately set to ensure sustainability and a profitable fishing industry. The same applies for water users – caps or limits on water use must be consistent with environmental sustainability. To accelerate autonomous adjustment with ITQs, whereby vessels that exit are bought by more-profitable operators, governments have initiated buybacks of fishing rights (Kompas 2006). When undertaking buybacks, explicit attention should be given to the differences between active and latent fishing licences; the risk is that inactive rights may be reactivated following the buyback. Australian governments are beginning to purchase water entitlements to meet environmental objectives, and to speed the transition to more efficient water outcomes. Such purchases would, however, need to take into account potential activation of water rights 'sleepers' (water rights that have never been used) and 'dozers' (water rights not currently used). Water tenders that force rights holders to undertake water efficiency improvements as a condition of sale will also raise the cost per litre of water purchased (Grafton & Hussey 2006).

3.2 Incentives

The key insight from fisheries is that price signals and incentives are critical to generating sustainable and profitable fisheries outcomes. In a review of over a dozen fisheries worldwide, Grafton et al. (2006) show that without appropriate incentives and adequate monitoring and enforcement, fishers frequently fail to adopt sustainable practices. The key is to ensure that fishers have durable property rights. The insight for water reform, recognised in the NWI, is the importance of creating durable property rights for water. In particular, holders of water rights should be able to trade their entitlements and seasonal allocations, the price of water should reflect the environmental costs of use, and non-compliance with regulations in terms of water withdrawals must be identified and remedied in a timely fashion.

The use of ITQs in fisheries also illustrates the need for ongoing management and innovation in terms of rights-based fishing. For instance, ITQs provide appropriate market signals for the use of target fish species, but do little to address concerns over bycatch of vulnerable non-target species, or activities that may impose environmental costs on others. Similarly, trading of water entitlements does not directly address issues of water quality, such as salinity. Additional policy instruments are required to account for water quality, such as appropriate environmental charges in terms of water quality or, as is currently done, restrictions can be imposed on trades that might reduce water quality (Heaney et al. 2006). The fundamental insight from fisheries is that operators need appropriate incentives to address water quality issues; when faced with such incentives they will act in ways to minimise these external costs.

The fisheries experience also stresses the importance of inclusivity of rights. Where ITQs have been applied and exemptions given, such as to operators of small vessels in Iceland, it encouraged large expansions in fishing effort in the exempted categories (Grafton et al. 2006a). In the case of water, metering and pricing of surface water use has sometimes encouraged greater withdrawals from groundwater sources, that may have detrimental impacts on downstream users and future water supplies.

3.3 Third-party effects

Where property rights are poorly specified, there are potentially unintended effects on other entitlement holders or the environment due to changes in water use decisions. Van Dijk et al. (2006) identify six factors that are poorly accounted for in property right specifications – long-term climate change, afforestation, groundwater extraction, changes to irrigation water management, farm dams and bushfires – all of which impose significant risks on the resource availability of water entitlement holders. Improvements in water accounting arrangements are needed so that a sufficiently comprehensive system can be developed to underpin water use and environmental water allocation decisions (PC 2006).

Heaney et al. (2006) identify a number of third-party effects related to water trade, where the costs or benefits of trading are not fully captured by holders of property rights. These include impacts on reliability of supply and delivery, storage and delivery charges, and water quality. These can be mitigated through more completely specified water rights although, in some cases, the costs of establishing property rights may be higher than the benefits they generate. In the case of fisheries, the concern has been that trades of harvesting rights will transfer fish from one region to another, with possibly negative social costs to some fishing communities. Some fishers have argued for restrictions on trade out of disadvantaged regions. The problem with this approach is that it freezes the current state of affairs and does not allow resources (fish) to go where they are most valued. Although superficially beneficial, trade restrictions provide no signal to resource users to adjust their behaviour and improve productivity, which is the ultimate promoter of economically viable communities.

The fisheries experiences with third-party effects offer insight for the trading of water. Constraints on water trading designed to protect communities or benefit remaining members of irrigation districts will, by themselves, fail to maintain viable communities. If fixed costs of supplying water are too high for the remaining operators after water trading, there are competitive pressures to change water use and how water is supplied. Maintaining the status quo via arbitrary trade restrictions or exit fees merely postpones the inevitable, and fails to give timely signals to water users to change existing practices.

4 Research gaps

To effectively implement the NWI a number of important research gaps must be addressed. First and foremost is a need for a better understanding of the ecosystem requirements of surface water to maintain desired environmental goals. This requires better knowledge of groundwater and surface linkages, the consequences of climate change, and the interactions between land use practices and their environmental impacts. This knowledge is required to manage risks (van Dijk et al. 2006) and to determine the appropriate caps and targets in terms of water use. To this end, at the summit on the Murray-Darling Basin it was agreed that CSIRO would report progressively by the end of 2007 on sustainable yields of surface and groundwater systems within the Murray-Darling Basin, including examining sustainable yield assumptions.

Better ecosystem knowledge and indicators of river health will provide useful information for valuing non-market benefits of environmental flows. These non-market benefits, at relevant spatial scales, provide the basis for pricing water not currently in use and the value of trade-offs between water withdrawals and environmental flows. At present these trade-offs are only implicit, determined by planning processes in terms of 'set-asides' of water for environmental flows.

Improved water outcomes for Australia greatly depend on the ability to support cost-effective water trading. Important research gaps include lack of knowledge on water prices, the transaction costs of water trades and the potential consequences of trades on communities. Research of these knowledge gaps will help to identify what policies may be needed to promote trades that contribute to efficiency, help resolve third-party effects and mitigate the social and environmental impacts of water trading.

Conclusion

This review of existing water trading and water pricing in Australia identifies key issues in the ongoing water reforms initiated as part of the NWI. First, it supports the NWI agenda to remove many of the restrictions on water trading. Second, it stresses the importance of water prices (including non-market values) and full-cost pricing as signalling the relative scarcity and value of water. Third, it emphasises the need to reduce reliance on inefficient regulations and controls to manage demand.

Much can be achieved in the coming years. The success of the water reform process, however, will be constrained by a lack of knowledge of the effects of water extractions on system-wide and local environments, non-market values of environmental flows, easily accessible and reliable water trading and pricing data, and an understanding of the social and economic impacts of trades.

References

Allen Consulting Group (2006). *Transaction costs of water markets and environmental policy instruments.* Water Study Report. Productivity Commission: Melbourne.

Australian Bureau of Statistics and Productivity Commission (2006). *Characteristics of Australia's irrigated farms: 2000–01 to 2003–04.* Catalogue No. 4623: Canberra.

ACIL Tasman in association with Freehills (2004). *An effective system of defining water property titles.* Land & Water Australia Research Report: Canberra.

Council of Australian Governments (2004). *Intergovernmental agreement on a national water initiative.* Council of Australian Governments: Canberra. Available at <http://www. coag.gov.au/meetings/250604/iga_national_water_initiative.pdf>.

Connell D, Dovers S & Grafton Q (2005). A critical analysis of the National Water Initiative. *Australasian Journal of Natural Resources Law and Policy* **10** (1), 81–107.

Crase L, O'Reilly L & Dollery B (2000). Water markets as a vehicle for water reform: the case of New South Wales. *Australian Journal of Agricultural and Resource Economics* **44** (2), 299–321.

Dwyer G, Loke P, Appels D, Stone S & Peterson D (2005). *Integrating rural and urban water markets in south east Australia: preliminary analysis.* Paper presented at Workshop on agriculture and water: sustainability, markets and policies, 14–18 November. Organisation for Economic Co-operation and Development: Adelaide.

Dwyer G, Douglas R, Peterson D, Chong J & Maddern K (2006). Irrigation externalities: pricing and charges. Productivity Commission, Staff Working Paper, March. Melbourne. Available at <http://www.pc.gov.au/research/swp/irrigationexternalities/index.html>.

Fenton M (2006). *The social implications of permanent water trading in the Loddon–Campaspe irrigation region of northern Victoria.* Report prepared for North Central Catchment Management Authority: Huntley, Victoria. Available at <http://www.nccma. vic.gov.au/Reports.asp>.

Grafton RQ & Hussey K (2006). *Buying back the living Murray: at what price?* Economics and Environment Network Working Paper EEN0606. Australian National University: Canberra. Available at <http://een.anu.edu.au/paper.html>.

Grafton RQ & Kompas T (2007). Pricing Sydney Water. *Australian Journal of Agricultural and Resource Economics,* in press.

Grafton RQ, Arnason R, Bjorndal T, Campbell D, Campbell HF, Clark CW, Connor R, Dupont DP, Hannesson R, Hilborn R, Kirkley JE, Kompas T, Lane DE, Munro GR, Pascoe S, Squires D, Steinshamn SI, Turris BR & Weninger Q (2006). Incentive-based approaches to sustainable fisheries. *Canadian Journal of Fisheries and Aquatic Sciences* **63** (3), 699–710.

Grafton RQ, Kompas T, McCloughlin R & Rayns N (2007). Benchmarking for fisheries governance. *Marine Policy* **31**, 470–479.

Hafi A, Beare S, Heaney A & Page S (2005). Water options for environmental flows. Australian Bureau of Agriculture and Resource Economics, Report 05.12. Prepared for Natural Resource Management Division; Department of Agriculture, Fisheries and Forestry: Canberra.

Hatton MacDonald D (2004). The economics of water: taking full account of first use, reuse and return to the environment. Client Report March 2004. CSIRO Land and Water: Canberra.

Heaney A, Thorpe S, Klijn N, Beare S & Want S (2004). *Water charges and interregional trade in the southern Murray Darling Basin.* Paper presented at 'Establishing water markets symposium', 9 August. Melbourne.

Heaney A, Dwyer G, Beare S, Peterson D & Pechey L (2006). Third-party effects of water trading and potential policy responses. *Australian Journal of Agricultural and Resource Economics* **50**, 277–293.

Kompas T (2006). Getting things right: structural adjustment in Australia's Commonwealth fisheries. In *Policy briefs: fishing futures* (eds RQ Grafton, T Kompas & K Barclay), pp. 8–9. Crawford School of Economics and Government, Australian National University: Canberra. Available at <http://www.crawford.anu.edu.au/pdf/policy_briefs/fishing_futures.pdf>.

McCann L & Easter LK (2004). A framework for estimating the transaction costs of alternative mechanisms for water exchange and allocation. *Water Resources Research* **40**, W09S09, doi:10.1029/2003WR002830.

Peterson D, Dwyer G, Appels D & Fry JM (2004). *Modelling water trade in the southern Murray-Darling Basin.* Staff Working Paper, November. Productivity Commission: Melbourne.

Peterson D, Dwyer G, Appels D & Fry JM (2005). Water trade in the southern Murray-Darling Basin. *Economic Record* **81**, special issue, S115–127.

Price Waterhouse Coopers (2006). *National Water Initiative water trading study.* Final Report. Department of Prime Minister and Cabinet: Canberra. Available at <http://www.dpmc.gov.au/nwi/docs/nwi_wts_full_report.pdf>.

Productivity Commission (2003). *Water rights arrangements in Australia and overseas.* Commission Research Paper, October. Productivity Commission: Melbourne.

Productivity Commission (2006). *Rural water use and the environment: the role of market mechanisms.* Research Report, August. Productivity Commission: Melbourne.

Quiggin J (2001). Environmental economics and the Murray-Darling river system. *Australian Journal of Agricultural and Resource Economics* **45**, 67–94.

Quiggin J (2006). Urban water supply in Australia: the option of diverting water from irrigation. *Public Policy* **1** (1), 14–22.

Qureshi ME, Proctor W & Young M (2005). *Economic impacts of water resources management in the Murray Darling Basin, Australia*. Paper presented in the XII World Water Congress 'Water for sustainable development: towards innovative solutions', 22–25 November. New Delhi.

Sibley H (2006). Urban water pricing. *Agenda* **13** (1), 17–30.

Van Dijk A, Evans R, Hairsine P, Kahn S, Nathan R, Paydar Z, Viney N & Zhang L (2006). *Risks to shared water resources of the Murray-Darling Basin*. Prepared by CSIRO for the Murray-Darling Basin Commission: Canberra.

Young MD, Proctor W, Qureshi MJ & Wittwer G (2006). *Without water: the economics of supplying water to 5 million more Australians*. CSIRO Land and Water & Centre for Policy Studies, Monash University: Melbourne.

Endnotes

1 Authorship is alphabetical. The views expressed in this chapter are those of the authors and do not necessarily reflect those of the Productivity Commission.

2 Trade in water entitlements, sometimes called 'permanent trade', involves moving the property right of a water entitlement in perpetuity. Trade in water allocations, sometimes called 'temporary trade', involves moving water allocations on a short-term basis, usually within an irrigation season.

3 Some of these derivatives can also have financial management and taxation benefits, compared with water entitlements.

Chapter 7

Linking rural and urban water systems

Mike Young[1]

Australia's climate and population growth and its National Water Initiative (NWI) are changing the way water is used in Australia. South-west, southern and eastern Australia appear to have returned to a drier rainfall regime reminiscent of the first rather than the second half of last century. Less rainfall means that less water runs into dams, aquifers and streams and rivers that we rely upon (Wentworth Group 2006). Furthermore, less water means that a larger proportion must be devoted to system maintenance.[2] To make the situation even more challenging, through the NWI governments have collectively committed themselves to restore health to overallocated surface and groundwater systems, *and* our population is increasing. Australia has a problem: in a time when it is looking for more water, it has to get by with less water.

With a focus on water use, this chapter explores the policy interface between urban and rural Australia. 'Urban' is interpreted to include industrial and commercial use as well as household use. The main finding is that policy choices at this interface will determine futures for people on both sides. Another is that some choices can be used to ease adjustment pressures and produce more efficient and arguably more equitable outcomes.

Historically, urban and rural water were managed separately. Typically, cities were built close to areas with sufficient water for their needs (Buxton et al. 2006). As populations have increased, however, cities have had to compete for access to water resources over greater distances: first in the peri-urban areas that surround them and, more recently, in all the water systems that can be connected to them.

We need to recognise that most water in Australia (as elsewhere) is consumed and lost in the pursuit of agricultural production. In the parts where urban and rural Australia co-exist, supply preference (priority) was given to urban needs. Essentially, these parts of the economy took what was needed and the environment and rural water users were left the residual.[3] Since the COAG water reforms of 1993/94 and, more recently, the 2004 NWI a progressive shift to a new regime has started. Full implementation of the NWI will mean that any increase in urban water use can come either from the development of new sources – desalination, recycling or stormwater capture – or by buying it from rural water users where that is possible. Under the NWI (para. 23), Australia is now committed to the use of market mechanisms to redistribute water across the nation as economic, social and environmental conditions change.

While the NWI (para. 64) proposes to 'facilitate the efficient functioning of water markets, including inter-jurisdictional water markets, and in both rural and urban settings' there is a big difference between facilitation and implementation. Governments remain reluctant to implement that which they have agreed to. In rural NSW, for example, town water entitlements are volumetric and determined by reference to the town's population. That is, rural town water entitlements increase as the town population increases and are not defined as a share of the available resource. In other parts of Australia, it has been traditional to license rural towns to simply take as much water as they need for reasonable domestic purposes. More recently, however, and as irrigators have started to appreciate the consequences of increased urban use, policies have been changed to require the purchase of water entitlements and/or allocations[4] from existing water users.

One of the advantages of trade is that while water transfers one way, money transfers the other way. Moreover, the process involves willing buyers and willing sellers. However, consideration of trade alone does not reveal the complete picture. While most trade occurs within the agricultural sector from farmer to farmer, the direction of nearly all rural–urban water trade is uni-directional from rural to urban systems,[5] although considerable financial transfers also move in the opposite direction. For example, this includes the considerable investment that urban Australians are making to restore rural environments, particularly the health of waterways and aquifers.[6]

Consideration also needs to be given to the extent that urban–rural trade affects environmental externalities and the interests of third parties. However, for the purposes of this chapter, and as other chapters attend to these issues in considerable detail, it is assumed that full implementation of the NWI coupled with the development of all the opportunities identified in this chapter will result in a framework where transfers of water and money between rural and urban Australia will always result in a nett benefit (when measured carefully and objectively against the counter-factual, i.e. what would have happened if transfers had not been allowed to occur?). This includes the introduction of complete accounting processes that recognise that many investments in so-called water use efficiency actually reduce returns back to the system so that, in terms of an aggregate improvement in water flow, money is not being spent for nothing.

1 Current knowledge and capacity

1.1 The value of water to the economy

Water is an essential input into many economic processes. While there is considerable opposition to the transfer of water from rural to urban Australia, formal economic analyses suggest that, at the national level, the transfer of water from rural to urban areas will increase GDP (Dwyer et al. 2005; Young et al. 2006). Water also contributes to the economy in other ways; one of the more notable is causing tension – the generation of hydropower. Water for the generation of hydroelectricity is most valuable in heatwaves when demand for electricity to run air-conditioning plants in urban Australia suddenly surges. This means that Snowy Hydro has an interest in leaving water in its dams until the surge occurs. Irrigators, however, want surety of supply that can only be given after the water has passed through a generator.

ABS data suggest that if the current suite of immigration and economic development policies remain roughly the same, over the next 25 years to 2032 Australia's population will increase by some 5 million people. Nearly all this expansion will occur in urban areas (see Figure 7.1). While 5 million people is a 25% increase in the population of Australia, the amount of water they can be expected to source from rural Australia is relatively small. Assuming that the available supply decreases by 15% in southern and eastern Australia and that all impediments to trading are removed, Young et al. (2006) have estimated that the volume of water they would seek to acquire from rural Australia – if allowed to do it via a market process – is less than 1% of the water extracted by all Australian water users. In total, around 61 GL of water is expected to transfer to urban households and 171 GL to commercial and industrial uses. The total average volume of water used for consumptive purposes in Australia is around 25 000 GL (ABS 2000).

While the transfer of water from rural to urban Australia can be expected to add more value to the economy and national well-being than leaving it in the rural sector, it is important

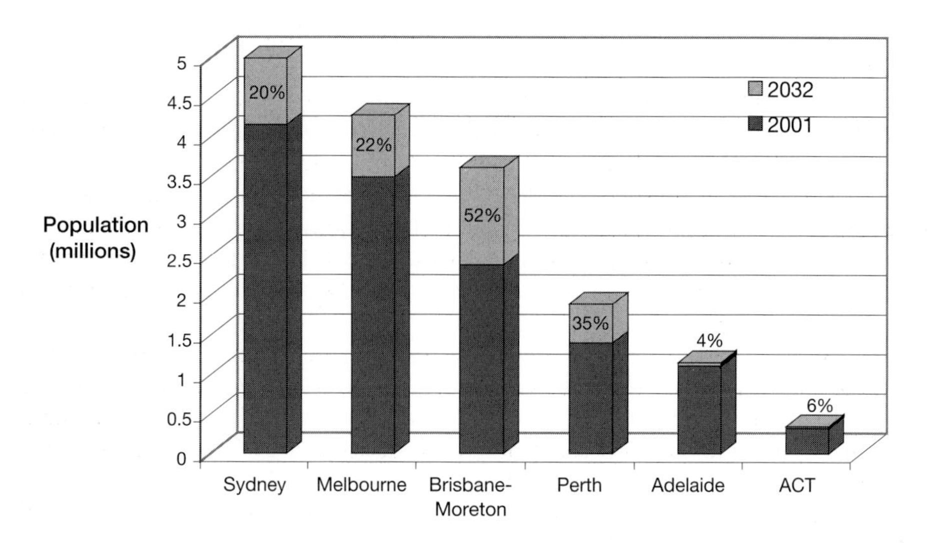

Figure 7.1 Indicative population projections for Australia's major urban centres.
Source: Young et al. (2006)

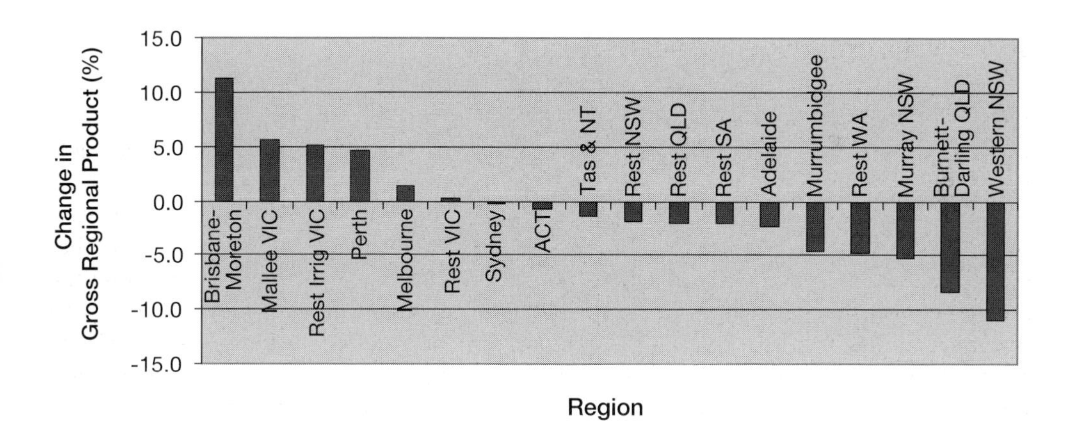

Figure 7.2 Percentage changes in gross regional product over 25 years as a result of removing all restrictions on rural–urban water trade.
Source: Young et al. (2006)

to acknowledge that the introduction of unrestricted urban–rural trading could have significant social and economic implications at the regional level (see Figure 7.2). Gains and losses are not uniformly distributed. Counter-intuitively, some rural areas can also gain from expanded rural–urban trade and some urban areas lose! These shifts are due, among other things, to the effects of competition between regions. Adelaide, for example, is already connected to and sources significant volumes of water from the Murray-Darling system. When (if) Melbourne connects to the Murray River system by accessing water in Lake Eildon at the top of the Goulburn River, jobs shift from Adelaide to Melbourne. Under the scenarios described above, in 2032 unrestricted rural–urban trading reduces gross regional product in Adelaide by 2.3% and employment in Adelaide by 1.8% (Young et al. 2006).

It is also worth noting that as urban populations increase, the volume of water consumed in the commercial and industrial sector is nearly three times the increase in household use (the homes that these people will occupy). As a general rule, restrictions on urban–rural trade mean that several urban and industrial jobs are lost for every job retained in a rural area. Adelaide's real Gross Regional Product drops by 2.3% and employment 1.8% as a result of competition for access to water. Adelaide is already connected to and buying water from the Murray River system for urban and industrial purposes. In urban areas, relatively small amounts of water can add a considerably large value to the urban economy at less short-term cost to the rural economy and the environment than most people appreciate (Dwyer et al. 2005). The main reason is that the volumes of water needed to generate $1 of urban income or value are much less than those in rural Australia. As Quiggin (2006) observes, water is typically sold to irrigators for around 10c per kL but after pressurising and purifying it sells for well over 100c per kL to urban water users.

In addition to considering implications for the development of Australian markets, it is also necessary to consider the nature of non-market values associated with water use. While there has been considerable analysis of the non-market values associated with actions that improve environmental amenity in our river and aquifer systems, there has been very little assessment of the non-market values that derive from the allocation of water to parks, gardens and lakes in urban areas.

Consideration also needs to be given to the distribution of externalities – effects on third parties and the environment – that are not resolved (see Gardner et al. 2006). The above economic analysis of the implications of Australia's increasing urban population assumes that many of the issues at the heart of the NWI are resolved according to plan. This includes the return of all surface and groundwater systems to health, and the resolution of some serious interception problems and failure to account for surface–groundwater interaction in a rigorous manner.

There is also a case for more research on the effects of infrastructure pricing and development policies on the nature of exchange between rural and urban Australia in the water sector. Removing barriers to water trade would provide clear price, opportunity cost and investment signals to all parties, but if investment in the infrastructure used to make water available for trade is irrationally priced, the resulting trade may be sub-optimal. When inefficient investments in infrastructure are made, these inefficiencies can be locked into the economy for a very long period.

Finally, there may be considerable unintended wealth transfers, which may be much more important than many appreciate. In a penetrating examination of urban water policies, Dwyer (2006) points out that much of the infrastructure used to supply water to Australia was not paid for by the people who use it. He goes on to point out that the accounting policies used to report on the 'profitability' of water utilities are very different from those used by private enterprise. As a result, the profitability of these businesses is seriously understated.

2 Adaptable knowledge

2.1 Pricing and other demand management issues

Among other things, the NWI aims to 'create policy settings which facilitate water use efficiency and innovation in urban and rural areas'. Conventional economic wisdom suggests that necessary conditions for such a policy setting include:

- removing all administrative impediments to trade in entitlements and allocations[7]
- pricing policies that subject both rural and urban water users and infrastructure managers to the same price disciplines.

Consideration of infrastructure pricing and charging policies is important as in both urban and rural systems the cost of securing access to infrastructure and using it is significant. While the NWI takes a strong view on removing administrative impediments within agricultural and environmental supply systems, it does not require all users (rural, urban and environmental) to be subject to identical price disciplines associated with water supply and is even less strict on the degree to which the cost of investments in infrastructure are to be recovered. Moreover, there is little detail as to the most appropriate way to encourage trade between urban and rural Australia.

One interpretation of the NWI is that, while the cost of water supply infrastructure is to be recovered via a range of upper- and lower-bound pricing measures, it does not require jurisdictions to recover the cost of grants made for indirect measures such as sewage recycling, replacing inefficient showerheads, installing rainwater tanks etc. In fact, the objectives of the NWI and the Australian Government Water Fund conflict. While most of the money placed in the Water Fund is being used to subsidise investment in water saving technology, the NWI sets out to use efficient market processes to achieve the same outcome. According to the National Water Commission, under the heading 'Best Practice Water Pricing,' the NWI sets out to:

> promote economically efficient and sustainable use of water resources, water infrastructure assets, and government resources; ensure sufficient revenue streams to allow efficient delivery of services; facilitate the efficient functioning of water markets; give effect to the principles of consumption-based pricing and full cost recovery; and provide appropriate mechanisms for the release of unallocated water.[8]

If all water users and all communities were expected to pay for the full cost of infrastructure projects and other activities, we would expect them to be very wary about applying for grants under the NWI's Smart Water program – any efficiency test of the kind required under the NWI would require the grants to be repaid. In particular, we would be surprised to see any proposals source water at much more than the known long-run marginal cost. This, however, is not the case and many projects under the NWI Smart Water program are being implemented at many times the cost of the most competitive source of water, namely, the acquisition of water via the marketplace. As a result, the market for infrastructure is being distorted and, arguably, the uncertainty around funding arrangements is discouraging the private search for innovative solutions to Australia's water management problems. Quiggin (2006) observes that Australian cities are facing severe, and in most cases chronic, shortages of water, yet as Dwyer (2006) points out private industry is not rushing in to invest. Further analysis of the impediments to resolution of this supply management problem is likely to produce significant dividends. Focusing upon the extent of the opportunities associated with this issue, the Business Council of Australia (2006) concludes that 'Australia's water problems are a direct result of a poorly planned and managed water system that has conspired to turn a sufficient supply of water at the source to scarcity for end users'. One of the many opportunities identified is the development of administrative arrangements to facilitate rural–urban water trade.

As a general rule, we would expect the real value of these subsidies to be capitalised into land and water entitlement markets. Recovery of the costs of Smart Water grants, possibly by converting them into a loan, would make it easier to achieve consistency with section 66(ii) of the NWI, which for metropolitan water requires 'the development of pricing policies for recycled water and stormwater that are congruent with pricing policies for potable water, and stimulate efficient water use *no matter what the source*, by 2006' (italics added).

Another source for confusion in the water market arises from differences in urban and rural pricing policies. While urban water utilities are required to continually move 'towards upper-bound pricing by 2008,' prices in rural areas need only achieve lower-bound pricing with continued movement towards upper-bound pricing 'where practicable'. At present, the size of this pricing gap is considerable. IPART's recent decision to require NSW to move to upper-bound pricing for rural users is estimated to collect an additional $30 million from irrigators. One interpretation of this policy might conclude that the pricing policies imposed on urban water users are stricter than rural water users. It is, however, also important to recognise that the rural water price is a supply price and that rural water users have the value or opportunity cost of scarce water resources revealed by the water market. In contrast, scarcity in an urban setting is revealed primarily through the use of regulations. As Table 7.1 indicates, recent research suggests that all of Australia's capital cities still subsidize water use.

Table 7.1 Revenue recovery in selected Australian cities

City	Current degree of full cost recovery	Revenue increase for 100% cost recovery
Sydney	76%	32%
Melbourne	68%	48%
Brisbane	84%	19%
Adelaide	82%	22%
Perth	74%	34%
Hobart	89%	13%
Darwin	80%	25%
Canberra	80%	25%
Total	75%	33%

Source: Marsden and Pickering (2006)

2.2 Extending the water market to include urban Australia

In practice, while a few urban water utilities have begun to buy water from people holding it for agricultural purposes, it is not yet possible for most commercial, industrial or household water users to buy and sell water – even if they are willing to pay a very high price for it. As yet, no independent analyst has assessed the cost of water restrictions on urban and rural water users but it is probable that the cost is considerable and its effects may be quite inequitable.[9] Arguably, rural water users also may be more exposed to the cost of externalities than are urban water users, as they are typically subject to more regional land use controls and planning processes.

Untangling the aggregate result of these competing considerations is probably extremely location- and industry specific, very complex and beyond the scope of this chapter. What it does suggest, however, is that Australia is still a long way from economically efficient national investment in water infrastructure and water use. One reason is the fact that Australian water policy has become inextricably linked to the delivery of political objectives. Arguably, one of the main reasons for the emergence of barriers to water trade as a vehicle for delivering local equity objectives, is that water trading restrictions remain one of the few instruments of government policy that are easily managed in a context-specific manner. While this may be the case, as a vehicle designed for the efficient delivery of social equity objectives, it is hard to think of a more inefficient means of doing so.

3 Alternative supply opportunities

As already noted, one of the consequences of distortions across the water market and infrastructure investment is that development of alternative supply opportunities in urban areas is perverse. In particular, the extent of urban water policies and the presence of barriers to urban–rural trade may be encouraging the development of alternative supply sources that otherwise would not be pursued. Wherever there are physical links between urban and rural water supplies, changes in one policy arena affect the other. In particular, the development of alternative urban water sources means that the demand for rural water is less. As Table 7.2 indicates, Australian experience with the cost of developing alternative sources of urban water suggests that costs are high, and in many cases considerably higher than those paid for access to water in rural areas (typically less than 20c per kL plus the cost of pumping water). This indicates that there may be considerable opportunities for the development of joint ventures and long-term supply contracts between rural and urban Australia. Where connection is possible, few technologies are cheaper.

Table 7.2 Comparative costs of accessing new water

Option	Total water supplied/saved (GL pa)	Cost/kL ($/kL)
Build 60 km pipeline and lift water metres	90	$0.40–0.50[d]
Appliance standards and labelling[a]	15	$0.04–0.05
Residential outdoor (excluding rain tanks)[a]	22	$0.10–0.20
Pressure and leakage reduction[a]	30	$0.20
Non-residential[a]	36	$0.30–0.50
Residential indoor – retrofits and rebates[a]	12	$0.50–0.60
500 ML/day desalination[b]	182	$1.73–1.98
500 ML/day indirect potable recycling[b]	182	$2.23–2.61
Committed/approved recycling schemes[a]	28	$1.00–3.00
Rainwater tank rebates – residential and schools[a]	2	$3.00
BASIX[a]	23	$0.30–4.00
Western Sydney recycled water initiative[a,c]	27	$5.80

Notes: Adapted from information collected by Marsden and Pickering (2006) and other sources.
(a) Institute for Sustainable Futures, ACILTasman and SMEC (April 2006), *Review of the Metropolitan Water Plan: Final Report.*
(b) *Sydney water: indirect potable recycling and desalination – a cost comparison.* Available at www.sydneywater.com.au. Accessed 24 July 2006. A smaller, 125 ML/d has also been considered – details were unavailable.
(c) Volume confirmed by Institute for Sustainable Futures (pers. comm., 8 August 2006).
(d) Estimated cost of pumping water from Murray Bridge to Adelaide.

3.1 Trading and infrastructure sharing opportunities

It is clear that there are significant opportunities for the use of market mechanisms to remove divisive them-versus-us language from Australian water use, but opposition to trading remains. A consortium of Victorian farmers has published a book expressing the view that the development of water markets that include urban water users and the environment will leave farmers 'high and dry' (Byrne et al. 2006). Among other things, the book recommends ongoing government investment in rural Australia with the immediate 'quarantine of irrigation water so that it cannot be permanently traded for urban, environmental or other issues so as to ensure security of supply to the agricultural sector'. Opposition to economically efficient water policies can also be found in the form of the NSW government's refusal to consider the transfer of water from agricultural use in northern NSW to the Gold Coast and, similarly, the Victorian government's refusal to consider the transfer of water from Victoria's Lake Eildon to Melbourne's water supply system.

However, elsewhere in Australia examples of a high degree of cooperation can be found. Notable examples include

- the transfer of 45 000 ML of agricultural water from Harvey Water to Perth in return for the upgrade of infrastructure
- Coliban Water's careful evaluation of sourcing water for Bendigo by purchasing water entitlements from irrigators in the Campaspe system (GHD 2006).

Arguably, the most complex examples of rural–urban water cooperation can be found in SA, where for many years a high degree of cooperation has existed. Examples, all of which are managed by SA Water, include:

- the transfer of 15 000 ML pa of recycled water from Adelaide's main sewage treatment works at Bolivar to irrigators in the North Adelaide Plains[10]
- the provision of low-cost access to the Mannum–Adelaide pipeline in winter and other low-demand periods to allow irrigators to pump water from the Murray River to the Barossa Valley where Barossa Infrastructure Ltd supplies it to irrigators[11]
- the purchase of 25 GL of water from irrigators in the Lower Murray swamps for Adelaide and 10 GL for Murray River environmental flow enhancement.[12]

An interesting feature of SA Water's approach includes the purchase of water entitlements and a willingness to sell it back to the irrigation sector until it is needed. South Australian experience, however, does indicate a need to make considerable investments in infrastructure for such arrangements to be successful. In particular, it is necessary to build a large interconnected set of pipelines and associated storage infrastructure.

Under Arizona's Assured Water Supply program 'every developer is required to demonstrate an assured water supply that will be physically, legally and continuously available for the next 100 years before the developer can record plats or sell parcels'. If the developer cannot assure supply, development may continue but the likelihood of inadequate supply must be revealed to aspiring purchasers.[13]

4 Research gaps and opportunities

While there are signs of the emergence of seamless water marketing and infrastructure sharing arrangements in Australia, the question remains whether administrative barriers to trade between rural and urban areas should be reduced. More particularly, there is a question as to whether the development of efficient infrastructure access pricing policies would improve water use across the boundary between urban and rural Australia. Economic reasoning would suggest that there is much greater opportunity for the use of pricing, as well as regulations, to manage water use in urban, commercial and industrial settings. Economists routinely point out that, if a parent is willing to pay for the cost of water used to allow a child to play under a sprinkler or in a swimming pool, the efficient solution is to allow them to do so. Similarly, if during periods of extreme water scarcity this results in a contest between rural and urban jobs then it may be more efficient for all if the question is resolved through market mechanism rather than political or administrative processes. Relying solely on regulation as a means of ensuring efficient water use is rarely likely to produce an efficient outcome.

One of the most undeveloped opportunities to reveal the extent of the scarcity value of urban water is to allow individual household, commercial and industrial water users to participate in the market for water entitlements. The answer to the question of whether urban–rural trading at the household level can be introduced at reasonable cost depends upon the degree of innovation in this area. It seems plausible, however, that all large water users could be allowed to buy and sell water. Similarly, it seems plausible that urban water supply arrangements could follow the path of electricity and gas supply arrangements. If this is the case, then water could be supplied by a number of retailers who are engaged in water trading on behalf of their customers. One way of doing this would be to transfer bulk water entitlements from the water utilities that currently hold them to individual businesses and local government councils. Under such an arrangement, any large business or local government that wished to expand water use would have to choose between:

- buying a water entitlement from someone then contracting a utility to supply it to them
- paying a utility to both source the water and supply it to them.

Although it is a larger conceptual step, wherever household water use is metered, household water bills could be turned into a quasi-water entitlement and a pricing arrangement put in place where:

- the charge for supplying water below a predetermined threshold would be at the marginal cost of supply
- the charge for supplying water above the threshold would be at the marginal cost of supply plus the cost of paying someone else not to use that water (its scarcity value).

Households could then be given the opportunity to trade in the threshold price at which water is supplied to them. Once such an arrangement was in place, all households would have the opportunity to purchase increased access to 'low-cost' water and/or sell water savings to any aspiring water user – including any irrigator or the environment. Once such arrangements are in place and the market for water develops, opportunities for investment in the development of alternative water supplies could be expected to emerge. If this occurs, the result would be a decreased demand for water from rural areas. Building upon the US experience in Arizona, approval to subdivide an area of land could be made conditional upon the supply and transfer of a water entitlement to each house or to the water utility responsible for supplying water to the development.

Another frontier in the early stages of development is the opening of access to sewage and stormwater and the development of mechanisms that allow this water to be traded between urban and rural systems. Often such arrangements involve substitution through the market's so-called 'invisible hand' As is the case in Salisbury in SA, a firm that previously sourced water from the mains now sources it from a stormwater capture system; the result is that less water is needed in the city and hence less is sourced from the rural system. Conceptually, the establishment of mature markets for these alternative water resources could result in the emergence of a suite of seamless institutional arrangements that collectively result in the optimal use of water across the nation – in households, in swimming pools, in business, in waterways and aquifers, in mining, in agriculture and for environmental purposes.

Conclusion

The overall conclusion is that in recent times water management issues have expanded to include consideration of the interface between rural and urban Australia. Water management is now seen as a both an urban and a rural problem. It is clear that policy choices in urban Australia affect rural Australia and vice versa; increasingly, decisions taken in one sector are influencing conditions in the other. Under the NWI it is envisaged, ultimately, that urban and rural Australia water supplies will be subject to the same price disciplines and traded in interconnected markets. Urban and rural water will be one and the same. Some recent policy decisions are consistent with progress towards this objective, but others are in conflict with it.

Water markets are starting to encourage the transfer of water from rural to urban Australia and generate greater returns for the nation as a whole. Considerable amounts of money are transferring from urban to rural Australia as the nation attempts to fix past mistakes. As a result, the signals being given to all water users are inefficient. The most inconsistent signals and hence the greatest opportunity for further research lies in the provision of infrastructure development grants.

In the long run and if the NWI is implemented fully, we can imagine a regime where water markets respond continuously to new opportunities and changing conditions. In pursuit of efficient outcomes, water and money could move seamlessly between both the urban and the rural sector and in a similar way among other sectors. If transfers between rural and urban Australia are to be efficient, considerable changes will have to be made to infrastructure development policies. In particular, it will be necessary for Australia to phase out infrastructure grants and to expose urban households (along with commercial and industrial business, the mining industry and environmental water managers) to water markets whose prices vary with scarcity, reflect the full costs of supply and reflect the full cost of externalities. Considerable changes to pricing policies and trading policies will be necessary. Arguably, the biggest research opportunity in this area is to find and develop low-cost ways of exposing the urban sector to the water market.

References

Australian Bureau of Statistics (2000). *Water account for Australia 1993–94 to 1996–97*. ABS Publication 4610.0. ABS: Canberra.

Business Council of Australia (2006). *Water under pressure: Australia's man-made water scarcity and how to fix it*. Business Council of Australia.

Buxton M, Tieman G, Bekessy S, Budge T, Mercer D, Coote M & Morcombe J (2006). *Change and continuity in peri-urban Australia. State of the peri-urban regions: a review of the literature*. RMIT University: Melbourne.

Byrne P, Eagle N, McDonald D & O'Brien J (2006). *High and dry: how free trade in water will cripple Australian agriculture*. Freedom Publishing: Melbourne.

Dwyer T (2006). Urban water policy: in need of economics. *Agenda* **13** (1), 3–16.

Dwyer G, Loke P, Appels D, Stone S & Peterson D (2005). *Integrating rural and urban water markets in southeast Australia: preliminary analysis*. Paper presented at OECD 'Workshop on agriculture and water: sustainability, markets and policies', 14–18 November. Adelaide.

Gardner A, Hatton T, MacDonald D & Chung V (2006). Pricing water for environmental externalities in Western Australia. *Environment Planning and Law Journal* **23**, 309–326.

GHD (2006). *Goulburn–Campaspe link project*. Interim report to Coliban Water and Goulburn–Murray Water. May 2006. GHD: Melbourne.

Grafton RQ & Kompas T (2006). *Sydney water: pricing for sustainability*. Working Paper 06–10. Crawford School of Economics and Government, Australian National University: Canberra.

Keating M (2006). Water pricing and its availability. Address to Committee for Economic Development of Australia, 29 November, Sydney. Available at http://ceda.com.au/member/nnx/multimedia/download/keating_water_nsw_speech_20061129.pdf.

Marsden J & Pickering P (2006). *Securing Australia's urban water supplies: opportunities and impediments*. Discussion paper prepared for Department of Premier and Cabinet, Marsden Jacob Associates: Melbourne.

National Water Commission (2006). *Progress of the National Water Initiative: a report on progress to the Council of Australian Governments*. 1 June 2006. National Water Commission: Canberra.

Quiggin J (2006). Urban water supply in Australia: the option of diverting water from irrigation. *Public Policy* **1** (1), 14–22.

Wentworth Group of Concerned Scientists (2006). Australia's climate is changing Australia. Statement by the Wentworth Group of Concerned Scientists, Sydney.

Young MD, Proctor W, Qureshi MJ & Wittwer G (2006). *Without water: the economics of supplying water to 5 million more Australians*. CSIRO Land and Water & Centre for Policy Studies, Monash University: Melbourne.

Endnotes

1. This chapter is intended to create open and transparent public debate and guide informed thinking on an issue of increasing importance. It was written with assistance and constructive criticism from Vanessa Findlay, National Farmers' Federation, and from Claude Piccinin, Water Services Association of Australia, Melbourne, Victoria.
2. The term 'system maintenance' is used intentionally. It includes water necessary to flush out pollutants, for navigation and for environmental purposes.
3. Examples of the extent of this approach can be seen in the tolerance and latitude given to the ACT, which has been nominally allocated access to much more water than it needs, and a special arrangement for SA cities dependent upon access to Murray River water that gives access to a near-guaranteed five-year moving average supply.
4. Consistent with the NWI, in this chapter the allocations are defined as an amount of water that may be used within a period. An entitlement is a right to receive allocations in a manner consistent with the licence and plans used to define the entitlement.
5. The exception is the return of recycled sewage for irrigation. A considerable proportion of the water in the North Adelaide Plains, for example, is derived from the Bolivar Sewage Plant that treats much of Adelaide's sewage.
6. The total transfer includes significant proportions of money made available under the NWI ($2 billion), the National Heritage Trust ($2.7 billion), and the National Action Plan for Salinity and Water Quality ($1.4 billion).
7. Physical limits on trade, the use of exchange rates and entitlement conversion factors etc. would remain. Trade would be possible only between physically connected systems.
8. www.nwc.gov.au/nwi/indes.cfm as at 19/11/06.
9. Phone calls to lawn-mowing franchises in Adelaide suggest that around 20% of the city's lawns are mowed by contractors; if lawn-mowing were banned this would reduce turnover by in excess of $22 million pa. The trade-off between jobs preserved per ML of water retained in rural Australia and ML of water traded to urban Australia has not been estimated.
10. www.sawater.com.au/SAWater/Environment/EnvironmentImprovementProgram/Virginia+Pipeline+Scheme.htm. Accessed 19 November 2006.
11. www.www.sawater.com.au/SAWater/AboutUs/AboutSAWater/irrigatin+services.htm. Accessed 19 November 2006.
12. Gee D (2006). Adelaide case study. Unpublished presentation to IWA Dry Area Forum, September 2006. Canberra.
13. www.water.az.gov/Watermanagement/Content/OAAWS/default.htm. Accessed 19 November 2006.

Chapter 8

Integrated assessment of impacts of policy and water allocation changes across social, economic and environmental dimensions

Tony Jakeman, Rebecca Letcher and Serena Chen

Integrated assessment of impacts from changes in water allocation and policy is a key implementation challenge of the National Water Initiative (NWI) as articulated in the *Water Perspectives* document of Land and Water Australia (2006). Knowledge is not always brought together in ways that integrate the social, economic and biophysical dimensions even when these interactions are fundamental. A series of five workshops across Australia identified the research needs for integrated assessment of terrestrial, aquifer and riverine systems. The network behind those meetings (www.stars.net.au) argued that while substantial efforts are being made worldwide to address sustainability-related issues in catchments, the approach has been too fragmented and would benefit from more coordination of research and management. Typically, research and policy have focused on one part of a catchment, one issue in isolation from others, or a subset of impacts and related disciplines. Often there is little emphasis on socioeconomic aspects, engagement with managers and community has been patchy, and continuity lacking. These limitations have led to ineffective, inefficient and sometimes conflicting attempts to solve water resource issues.

Australia has made as much progress as any nation in water reform (see Chapter 11, this volume), clarifying the nature of rights in water, separating regulatory and supply functions, and providing for environmental flows. But progress is challenged in many catchments where overallocation is pervasive, farm dams, plantations and other uses are expanding to reduce water supply and groundwater use is not sustainable. 'The variability of our climate and the location of water resources compared with growing population centres and the importance of rural industries reliant on irrigation create a suite of circumstances particular to Australia' (Chapter 1, this volume).

The challenge is for effective water planning and management that: treats water as a single resource over whole catchments where the connectivity of land use and surface water, groundwater, lakes and floodplains is recognised; minimises the inequitable and inefficient distribution of resources; reduces vulnerability to excessive demands; and limits the impacts of water allocation on their quality and ecological function. Meeting this challenge demands frameworks and tools that connect these issues and, as far as possible, quantify the trade-offs to inform decisions. Without such a capacity, comprehensive and integrated water planning required under the NWI (Chapter 5, this volume) – especially that which accounts for other resources and issues – will be difficult to achieve.

The discipline of integrated assessment (IA) affords a promising vehicle for addressing interconnected issues surrounding water allocation and policy. It attempts to tackle real-world management issues by integrating multiple issues and impacts, with knowledge, tools and understanding from multiple research disciplines as well as from the community, industry and managers. The last dimension of IA – the integration of stakeholders – allows us to incorporate their needs, aspirations and knowledge so that the process is a collaborative one that is more informed and likely to lead to acceptable trade-offs and outcomes.

This chapter identifies available IA methods and approaches, and their relative strengths, weaknesses and suitability to NWI tasks. The aim is to facilitate more confident interrogation of research questions and the determination of more reliable answers. The chapter concentrates on broader integration methods rather than on specific subsidiary methods, and addresses some gaps and opportunities in the latter, more disciplinary areas.

1 Key features of IA

IA is defined as:

> Integration of knowledge from different disciplines with the goal to contribute to understanding and solving complex societal problems, that arise from the interaction between humans and the environment, and to contribute in this way to establishing the foundation for sustainable development. Modelling and participatory processes should include stakeholder groups and the public at large (www.tias-info.web).

The key features of IA summarised by Jakeman and Letcher (2003) are consistent with many of the rationales listed in Land and Water Australia (2006), as illustrated in the quotes from the latter document:

- a problem-focused, iterative and adaptive approach linking research to policy, recognising that 'there are some real challenges in defining the social and economic impact questions that are to be answered'
- interactive transparency and 'trustworthy methods' that enhance communication to 'balance, integrate or trade off environmental, social, cultural and economic values'
- enriched by stakeholder involvement and is dedicated to adoption
- takes into account, using 'reliable and rigorous tools', the key complexities between the natural and human environment, spatial dependence, feedbacks and impediments
- attempts to recognise missing essential knowledge and 'to clarify both the potential and limitations of tools and methods and (hence) the trust that policy-makers and the community can place in the data generated from them'.

Multifaceted integration needs a set of tools to be effective. One of these is integrated scenario modelling (ISM), where we trace the effects of controllable and uncontrollable drivers through a catchment system, in this case the system where water allocation and other policy is being considered, and calculate the effects on a range of socio-economic and environmental indicators (Figure 8.1). The system is represented as an integrated set of models, often a mix of quantitative and qualitative types, and may have been rigorously tested and/or be expert opinion. ISM can be an investigative tool which, in scanning alternative futures or policy options, can facilitate systems thinking and learning by user groups and lead to innovation. An example of this is given in Ticehurst et al. (2007), where an integrated model framework was developed to assess the sustainability of coastal lake-catchment systems in NSW. This framework allows the exploration of the effects of controllable drivers, such as farm management and land use change, and uncontrollable drivers, such as sea level rise.

The impacts of climate variability and change are a major concern in relation to water management, as they can affect water resource availability. IA can provide a means of identifying climate impacts on catchments and lakes, which include not only water availability but also effects on streamflow, agricultural water demand, erosion and crop yields etc. (see Merritt et al. 2004). The ability to explore and consider the effects of controllable and

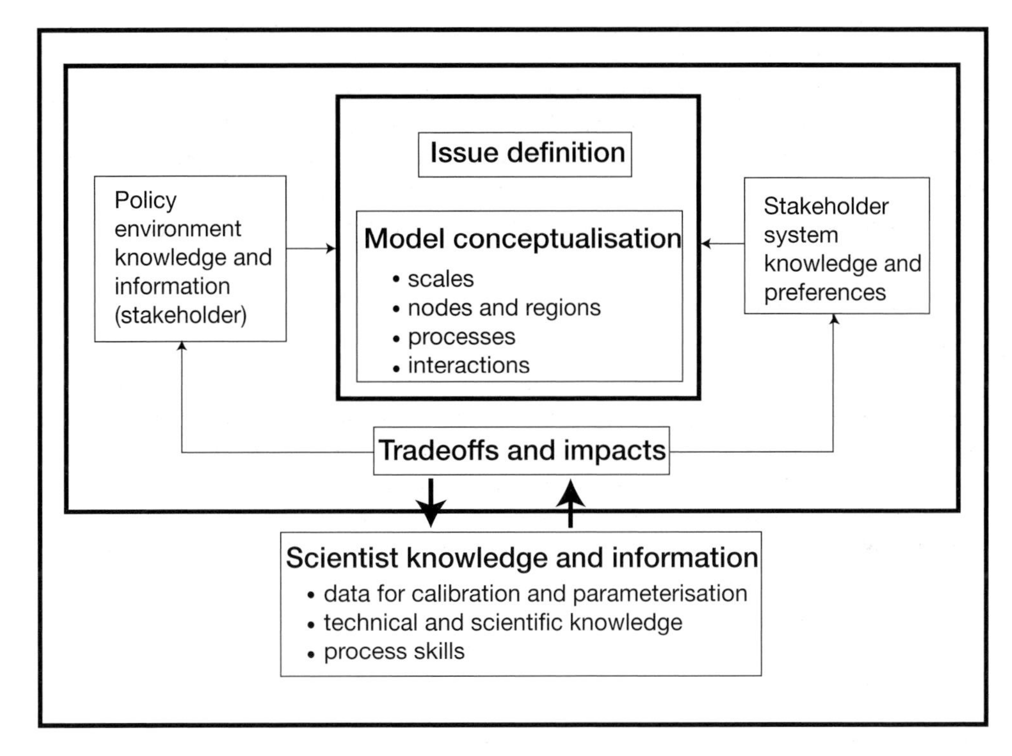

Figure 8.1 Diagrammatic representation of IA.

uncontrollable drivers, such as climate, is a key IA feature that allows for better-informed policy-making.

A second modelling approach utilises optimisation, where the model explicitly determines the best intervention or decision according to a specified objective (maximise net returns, minimise environmental costs etc.) subject to various constraints. The model user is generally presented with a single 'best' option or intervention. The objective function may be defined as a weighted combination of multiple outcomes.

Policies may be investigated within an IA as scenarios. These may be explicit drivers that allow us to directly interpret the way land and water is utilised by individual or industry decision-makers, or they may be aimed at changing the behaviour of communities or individuals in more subtle ways, such as through covenants or incentive schemes. Policies and other options that may be explored by IA include:

- instruments such as water markets, trading and pricing arrangements, regulation and information instruments, either individually or as a mix
- allocation rules
- access rules such as commence-to-pump thresholds, water sharing arrangements and carry-over limits
- infrastructure investments
- land zoning and development
- Indigenous and wider community access and use, including non-extractive uses of water
- interacting policies from non-water related agencies, such as the placement of roads and non-water related infrastructure, or health management and food safety and interactions with water quality

- adoption of new water technology
- remediation efforts such as riparian planting or engineering works and policies designed to increase their uptake
- placement and extent of plantation forestry
- water quality targets, e.g. pollution and salinity
- restrictions on commercial and non-commercial uses of water bodies and their flora and fauna.

Uncontrollable drivers that may be explored include:

- climate scenarios
- international agreements
- trade prices and international markets
- demographic changes
- technology availability
- community preferences.

Alternative or component tools that can be useful in any IA include:

- cost–benefit analyses
- multicriteria analyses
- risk assessment methods
- optimisation tools
- scenario tools
- visualisation tools
- a wide range of disciplinary tools including those in hydrology, ecology, social impact assessment and economics
- decision support systems.

2 The IA process

Various approaches have been developed for undertaking an IA (Lee 2006; Castelletti & Soncini-Sessa 2006; Harremoës & Turner 2001). These processes have a number of common features including a strong commitment to co-design of the assessment by stakeholders and inclusion of less certain or postulated processes that have been identified as important by stakeholder and expert consultation. This section briefly outlines one such process, developed as part of the Coastal Lakes Assessment and Management (CLAM) approach (Ticehurst et al. 2005) to illustrate the main features of an IA.

Step 1. Build understanding of constraints and issues in the catchment as well as possible targets and measures of catchment health. Also consider industries that are likely to be affected (positively and negatively) by changes and policies that operate and will affect impacts or actions. This step involves reviewing existing information on the catchment (management reports, previous studies etc.).

Step 2. Develop an initial conceptual framework, identifying key drivers such as management, development and remediation options as well as state and utility variables and their interactions. This framework is generally developed in-house using information from reviewing reports and other information available on the catchment. Discussion with key stakeholders would be expected at this stage.

Step 3. Workshop the initial conceptual framework and general scenarios with stakeholders to obtain feedback on missing scenarios, links and impacts. This involves broad consultation with many different stakeholder groups for feedback across social, economic and environmental issues. It includes discussion of specific scenarios to be considered (e.g. moving from 'urban development' to 'urban development in specific locations with specific controls'). It should consider the states that would be most usefully output from the model when users are considering the impacts. For example, the impact on total nitrogen delivered to an estuary may be expressed as a percentage change in total load or concentration, a frequency of exceeding a standard or the probability of achieving some other catchment target. Stakeholder feedback is essential to ensure that model outputs are useful for decision-making and other intended uses.

Step 4. Revise the initial framework using stakeholder feedback obtained in Step 3, producing a working version of the framework for the IA and identifying specific scenarios and key impacts.

Step 5. Identify existing data and information to populate the assessment, including datasets or other models (e.g. water quality or estuarine models), which may be used within the IA. Assess the way in which this information may need to be modified or used within the IA. For example, a model may need to be run thousands of times under different assumptions to provide the information required by the IA. Alternatively, a model may inform expert opinion if it is considered to be lacking key processes or information.

Step 6. Identify and fill key knowledge or information gaps. This step involves reviewing the working conceptual framework to identify links and utility variables for which no or very limited information exists, then constructing a work plan to fill these links and variables with information. If funds are insufficient to fill all gaps, they can be prioritised and non-priority links may be populated with expert opinion that reflects the high level of uncertainty in that information.

Step 7. Populate the IA model, running models or analysing data to produce the required information. This step is time-consuming, relying on a vast number of model runs and a high level of analysis. This analysis needs to be informed by the requirements of stakeholders identified in Step 3 with regard to the states used to describe model outputs.

Step 8. Review the assessment information using expert and stakeholder feedback. All assumptions should be carefully documented and ideally should be available from within any model platform. They need to be reviewed and comments on their validity at least documented and, where resources are available, acted upon.

Step 9. Incorporate the IA model in a software environment, typically including a user-friendly interface. This software should aid in the communication of concepts and science within the assessment and allow easy access to scenarios and results.

Step 10. Review the interface and populated IA model with stakeholders. The IA is demonstrated to stakeholders, seeking feedback on the usability of the system, explanation text provided in the interface, and the model results.

Step 11. Revise the interface and populated IA model, to incorporate stakeholder preferences and feedback.

Step12. Distribute the integrated catchment model to relevant stakeholders with appropriate training in its use, including the software system, model documentation and training materials. These are generally distributed through workshops run with people the client has identified as key users. They may include local or state government staff or consultants, regional organisations or community groups.

3 Integration methods in IA

3.1 Considerations in selecting a method

There are various approaches to integration and an increasing literature that uses these approaches to solve a wide variety of problems. Letcher and Jakeman (in press) categorise them into six major approaches that suit different application situations. As with all modelling problems, integrated model developers and users need to first have a good understanding of the needs of their model and of the types of data available before they select an approach. They must ascribe the reasons for modelling among priorities that mainly include prediction/ forecasting, system understanding, social learning and management/decision-making.

'Predictive models' are generally required to have reasonable accuracy in reproducing historic observations of system outputs from observed inputs. For integrative models, validating the predictive accuracy of these models can be difficult due to a lack of appropriate data for validation. Models for 'system understanding' tend to be research models, accessible to the model-builder and other researchers and technical specialists, as opposed to 'social learning models' developed for a non-technical audience. Those emphasise the importance of interactions between individuals or groups. The emphasis in models developed for social learning tends to fall on the plausibility of interactions and outcomes, rather than on predictive accuracy. 'Management' and 'decision-making models' usually have to be able to accurately differentiate between decisions or management options, requiring the model to give sufficiently accurate estimates of the magnitude and direction of changes in system outputs in response to changes in system drivers. Such differentiation or relativism means the models may have less-demanding requirements on accuracy than the predictive models that tend to be more concerned with absolutes.

The objectives of integrated modelling may involve one or more of these objectives, perhaps all, to some extent. The task is to set priorities among the objectives. Any framework for choosing an appropriate integrated modelling approach must also consider the spatial and temporal scales required, the degree of reliance on qualitative data, and characterisation of uncertainty. There will always be a compromise between representing depth in individual system components and breadth of the overall system. The challenge is to capture the advantages of these approaches while overcoming some of their limitations, possibly through the development of more hybrid models using a variety of approaches to model integration.

3.2 Five methods

There are several different approaches to integration; we discuss five major ones here. Some overlap in terms of definition or scope can be allowed in a combined approach. Bayesian networks, agent-based models and expert systems, described below, might be considered as artificial intelligence or knowledge-based techniques. Their individual use as integrating frameworks warrants particular mention. Table 8.1 summarises their properties.

System dynamics

System dynamics is a modelling approach that investigates and manages complex feedback systems. In theory, system dynamics approaches can be applied to various situations such as decision-making and policy design. In practice, this approach is most commonly used to increase system understanding and experiment with alternative system assumptions or for social learning applications. This is because this approach generally involves developing models with complex feedbacks and inclusion of less well-known or understood processes through inclusion of possible model structures (i.e. processes and connections are linked by plausible connections).

Bayesian networks

Bayesian networks (BN) can be considered a fundamental modelling tool for decision-making and management where representation of uncertainty is a key consideration. These networks represent uncertainty through conditional probability distributions (Pearl 1990). BN are

Table 8.1 Summary of approaches to integration

Approach	Uncertainty in model structure	Uncertainty in data/input values	Space treatment	Time treatment	Data types (quantitive or qualitive)	Optimisation or scenario based	Applications
System dynamics	No – easier to test than others	No	Various	Various	Quantitative	Scenario	System understanding/ experimentation Social learning
Bayesian networks	No	Explicit	Various – lumped, non-spatial more common	Various – lumped temporal or non-temporal more common	Both	Scenario	Decision-making and management
Coupled component models	No	No	Various	Various	Quantitative	Both	Prediction, forecasting Decision-making and management System understanding/ experimentation Social learning
Agent-based models	No	No	Various	Various	Quantitative	Scenario	Social learning System understanding/ experimentation
Expert systems	No	Explicit	Various	Various – usually non-temporal but rules can be 'forecast' based	Both	Scenario	All

useful when observed data are qualitative or inadequate but other sources of information are available. Unlike other integrated modelling approaches, BN use probabilistic, instead of deterministic, methods to describe the relationship among variables and thus uncertainty can be represented. Bayesian decision networks are modified BNs that include decision (management) variables and utility (cost–benefit) variables. Relationships among variables (nodes) are represented by arrows showing causal dependencies or by an aggregate summary of complex associations. Thus nodes in the conceptual framework of a BD network represent state, decision, utility or impact variables, while arrows between nodes represent conditional probability distributions (see Ticehurst et al. 2007).

Box 1: Bayesian network for the Eastern Mancha region, Spain

Issues/problem: The Eastern Mancha region (8500 km^2), Castilla-La Mancha, Spain, is faced with the risk of overexploitation of its local aquifer due to the significant increase in irrigated arable land over the last 25 years. This overexploitation could cause serious environmental damage, including contributing to the desertification process, and lead to uneconomic outcomes in the future.

Stakeholders: Farmers, urban and industrial water users, local community, relevant resource managers.

Integration methods and models: Bayesian network consisting of 56 variables arranged into five groups: input volume, environmental restrictions, urban demand, agricultural demand and result variables. The agricultural demand group contained five subgroups: limitations of irrigation systems, maximum volume available for irrigation, agricultural income, exploitation plan control capacity and irrigation consumption.

Participatory mechanisms: Active stakeholder involvement was encouraged during the design, development and testing stages of the Bayesian networks. Prospective participants were sent a questionnaire which aimed to identify the user and their decision-making capacity, provide information on the unit and gather data about their activities and opinions about how to improve water resource management. After the model was built, it was tested through stakeholder meetings.

Outcomes: The model can help in the sustainable management of the Eastern Mancha aquifer by balancing the various demands against the volume of recharge per year. Of the three scenarios proposed, the one considered best was sustaining the current water use for irrigation, and increasing the surface water use and the irrigation system efficiency. The scenarios were affected by many variables, leading to a high level of uncertainty, which can be dealt with using Bayesian networks. The involvement of stakeholders through the different processes enhanced the acceptability of results to all water resource users.

Source: Integrated water resources management of the Hydrogeological Unit 'Eastern Mancha' using Bayesian belief networks. In Martín de Santa Olalla et al. (2005).

Coupled component models

This approach combines models from different disciplines to arrive at an integrated outcome. Coupling may be loose, where outputs from models are linked together manually or externally to the original models; or the coupling may be tight (e.g. Letcher et al. 2004, 2007) where component models share inputs and outputs. The conceptual framework for a coupled component model generally represents links between system components, so that nodes often represent detailed component models and links correspond to data passing between models. These models are generally able to incorporate feedback. Coupled component models suit problems that require non-trivial temporal and spatial aspects to be clearly discrete. They are also useful for many catchment problems that require the underpinning of a nodal network where we can assess the impacts of upstream nodes (and their subcatchments) on downstream nodes.

Box 2: Coupled component model for the Namoi River catchment, Australia

Issues/problem: The Namoi catchment (42 000 km^2) in northern NSW is faced with problems of overallocation of groundwater and surface water supplies. This is an important irrigation area, and managing the allocation of its water supplies has significant social, economic and environmental impacts. An integrated model was developed as a means of considering economic and environmental trade-offs of alternative water allocation options.

Stakeholders: Farmers, local community, various departmental office staff, members of the River Management Committees in the catchment.

Integration methods and models: Coupled component model consisting of an agricultural production (economic) model, hydrological model, policy model, and extraction model.

Participatory mechanisms: Stakeholders were involved from the initial stages of identifying management needs and concerns, and the alternative options available to catchment managers. Stakeholders also contributed in the model development process, where they queried and in many cases corrected modelling assumptions and identified alternative resource use and management scenarios. Final feedback was sought through a series of public seminars and discussions, which included detailed presentations on model structure, input data assumptions and results. This feedback from stakeholders influenced future modifications to the model.

Outcomes: The advantages of the model, according to stakeholders, included the integrated approach that assists in making better policy decisions, its ability to help clarify the relative impacts of changes at the large scale and convey information, its flexibility and accessibility, and the open (transparent) process of its development. Limitations identified by stakeholders included the lack of a groundwater modelling component, lack of links between climate and crop yields and water use, inaccurate assumptions about pumping flood flows and about the decision-making behaviour of farmers, and simplified representation of one subcatchment. The IA was a learning process for researchers and community members, and helped to improve communication and understanding between policy-makers, stakeholders and researchers.

Source: Application of an adaptive method for integrated assessment of water allocation issues in the Namoi River catchment, Australia. In Letcher & Jakeman (2003).

Box 3: Coupled component model for the Maipo River basin, Chile

Issues/problem: The Maipo River basin is located in an important agricultural region in the metropolitan area of central Chile. In the mid 1990s, agriculture accounted for 64% of the basin's total withdrawals, and the domestic and industrial sectors withdrew the remaining 24 and 11% respectively. Competition for water among the different sectors is rapidly increasing, presenting the challenge of appropriate management of the scarce resource.

Stakeholders: Agricultural, domestic and industrial sectors.

Integration methods and models: Integrated economic–hydrologic modelling framework, containing three main components: 1) hydrological (includes the water and salt balance in the basin's reservoirs, river reaches and aquifers), 2) water use (includes water for irrigation and municipal and industrial uses), 3) economic (includes sum of benefits from irrigation, hydropower and municipal/industrial demand sites). The model considers interactions between water allocation, farmer input choice, agricultural productivity, non-agricultural water demand and resource degradation. This provides a means of estimating the social and economic gains from improving water allocation and water use efficiency. Different demand management strategies were evaluated, including changes in water price and markets in tradable water rights.

Participatory mechanisms: Indirect.

Outcomes: The model can test different scenarios, with changes in variables such as hydrological levels, irrigation technology cost, crop price, source salinity, irrigated area and demand management strategies. Due to complex interactions within the system, these changes in variables can be reflected in water and salinity levels, food production, economic welfare and environmental consequences. The model results showed that water trading provided significant benefits, with water transferred to higher-valued agricultural and municipal/industrial uses. This model can assist policy analysis and water allocation. The authors identified additional future research needed, including the extension of the agricultural production functions to incorporate inputs such as chemicals and labour, refining the estimates of the urban water demand functions and power generation, and establishing direct cooperation with stakeholders.

Source: Integrated economic–hydrologic water modelling at the basin scale: the Maipo river basin. In Rosegrant et al. (2000).

Agent-based models

Agent- or actor-based models can be considered an example of coupled component models. They tend to be very hypothetical and are usually applied in social and ecological science. Agent-based models focus on representing interactions between individuals or agents in a system (Brown et al. 2004). An agent has the ability to adapt when its environment changes. Multi-agent systems are made up of two or more agents who exist at the same time, share the resources and communicate with each other. A key focus of agent-based modelling is the discovery of emergent behaviour, large-scale outcomes resulting from often-simple local interactions between individuals. Thus they are useful for social learning applications. As for coupled component models, the conceptual framework for an agent-based model usually demonstrates links between detailed component models.

Box 4: A multi-agent model for the Kairouan water table system, Tunisia

Issues/problem: The Kairouan water table, in Central Tunisia, has been decreasing for more than two decades as a result of overexploitation by private irrigators. The dynamics of this system were considered to be strongly influenced by local interactions between the resource and its users, and direct non-economic interaction between farmers. A multi-agent model was developed to represent the complex and distributed system, and explore the interaction between physical and socio-economic components, and to provide a tool to help authorities identify a socially acceptable approach to managing demand in order to protect the resource.

Stakeholders: Farmers.

Integration methods and models: SINUSE, a multi-agent model, was applied. It contained three types of entities: social entities (farmers), spatial entities (plots, public irrigated perimeters, water table zones) and located entities (wells and boreholes). These entities can interact, for example farmers can interact with one another in relation to the construction of wells and land exchanges. Certain requirements might be necessary before an action is carried out, for example to build a borehole a farmer needs at least 20 ha of land, a certain amount of money to cover its costs, and a risk-free location for the bore. The watertable level depends on hydrogeological parameters. The model was based on a series of assumptions. For example, all plots and farmers were assumed to achieve the same yield, except for dry farming and market gardening, which are influenced by climatic or economic variations.

Participatory mechanisms: Detailed inquiries were made on a selection of representative farmers over one year. Statistical inquiries were also made on 10% of farmers. These inquiries enabled the researchers to define the various components of water demand, and helped improve their knowledge about the farmers' behaviour (crop choice, investment decisions).

Outcomes: Local and non-economic behaviour, often now considered in other models, was found to have a major impact on the dynamics of the system. The development of this model improved knowledge of the system, especially in regard to understanding the factors determining water demand and farmer behaviour. It was suggested that this model could be a useful tool for negotiating the integrated management of the water table system due to its transparency and ability to simulate different scenarios. The authors acknowledged that a return to the field was necessary to verify socio-economic assumptions and to socially validate actor behaviour.

Source: SINUSE: a multi-agent model to negotiate water demand management on a free access water table. In Feuillette et al. (2003).

Expert systems

These are a type of qualitative model where knowledge is encoded into a knowledge base and the expert system uses logic to infer conclusions. The aim of an expert system is to simulate problem-solving tasks to tackle the problem at hand (Kidd 1987). They have the ability to learn from the experience of the user and knowledge inputs. The knowledge base determines the success of the system (Forsyth 1984). The conceptual diagram for an expert system refers to questions about the nature of the system directed at the user. The response to these questions then dictates the route down which the procedure looks for a solution.

Environmental informatics and artificial intelligence

The emerging field of environmental informatics has much promise, combining artificial intelligence, GIS, software frameworks for modelling and linking models, and user interfaces. However, effective capture of knowledge has lagged far behind software implementation. As with artificial intelligence, environmental informatics urgently needs to be complemented by user-driven IA providing priorities, structure and efficiency.

4 A role for IA in the NWI

Not only does the NWI invite some form of IA, it can be a valuable discipline in addressing many implementation challenges. It is clearly relevant for water planning, conceptualising and modelling the link between rural and urban systems, including Indigenous perspectives, informing audit and review processes, assessing the impacts of water markets, exploring options for environmental flows and supporting adaptive management. In particular, IA can assist with expressing and communicating around multiple values (see Chapter 2, this volume), through frameworks to inform planning that integrate social and cultural values, and structuring stakeholder involvement and understanding of the planning and other processes occurring under the NWI.

Water policy and management is a highly complex problem which requires the understanding and consideration of a diverse range of dimensions (social, economic and environmental) to be effective and well-received by stakeholders. This can be achieved through a participatory and collaborative process, as offered by IA.

5 Research gaps

We know some of the important information that needs to be gathered to progress water resource planning and management through IA. The social sciences can offer insight and information into decision-making and adoption processes previously ignored in many scenario-based models. In particular, social survey data, linking information about decision-making and adoption to the biophysical and socio-economic characteristics of farmers, industries or households, is crucial to developing more sophisticated IA and other policy analyses. Very little of this type of data exist for river basin situations. In addition, biophysical scientists are often not in a position to extract and understand the implications of such data.

Further use and development of participatory methods (Haslam et al. 2003) for ISM building is one way of extracting and using such information, with the bonus of including stakeholders in model development. This ensures that stakeholders have a better understanding of, and opportunity to feed into, the assumptions underlying these types of models. Hare et al. (2003) present a comparison of different participatory processes. Most IA approaches contribute to capacity-building, providing opportunities to increase community awareness and understanding of trade-offs and linkages in the catchment system.

Given the complexities and uncertainties of integrated modelling, it should be accepted that its broad objective is to increase understanding of the directions and magnitudes of change in impacts under different options. Typically, it cannot be about accepting or treating simulation outputs as accurate predictions. We need qualitative differentiation between outcomes, with at least qualitative confidence; for example, a particular set of impacts or indicator values might be categorised as overall better than, worse than or negligibly different from another set with high, moderate or low confidence. This can be enough to facilitate a decision on the worth of adopting a policy or controllable change. ISM analyses must be able to differentiate between policies and specify what knowledge or data will provide leverage to improve the differentiation. Ideally, predictions would be produced with a quantitative confidence level, but in most current situations this is impracticable. Currently, methods for quantifying uncertainties have limitations and research is required to tune uncertainty methods to the needs of integrated assessment.

Adaptive management needs more emphasis and strategic research. It can involve the development of ways to gather, record and share conventional and unconventional environmental system information; improved tools to capture and express qualitative knowledge; methods for testing knowledge, identifying gaps and designing experiments; development of monitoring techniques that can distinguish the effects of changed management practices from natural variations; approaches to screening and testing alternative policies; and incorporation of the principles of feedback control to achieve acceptable behaviour insensitive to disturbances and modelling error. Norton and Reckhow (2006) offer some ways forward, including having more systematic and consistent problem-solving procedures that are open to scrutiny, and performing periodic parameter updating of predictive models and revision of future planned action to prevent sustained discrepancies between model and system behaviour.

A final knowledge gap is catchment-specific conceptual frameworks that are developed in conjunction with managers and stakeholders. These are required for integrated assessment of impacts and can inform where to focus disciplinary research efforts to achieve the best management outcome in a catchment system. These frameworks need to cross issues and involve a broad variety of trade-offs, impacts and processes, and to be informed by community values and expert knowledge. They are essentially the first step in any IA process and provide a framework for later integration of disciplinary assessments.

Conclusions

The level of sophistication of an integrated assessment should depend on context – the questions it is addressing and the knowledge base and resources available. The overall trend should be toward more clever frameworks and tools that recognise context, rather than sophistication. On the positive side, a range of integration frameworks already exists to help address the environmental, social and economic impacts of water allocation and other policy instruments, combining integrative capacity and a practical, participatory style of investigation and problem-solving. Their potential to clarify options and generate partnerships has already been demonstrated in various case studies (Giupponi et al. 2006; Jakeman et al. 2005) but we still need to further develop them by adapting them to new case studies tuned to Australia's water reform agenda. Different approaches are suited to different situations, and the characteristics summarised in Table 8.1 are one means of informing choice of approach. The participatory potential of IA approaches should be recognised, and again close collaboration with managers and stakeholders in the planning process allows the most appropriate IA to be delivered (see Table 8.2).

Table 8.2 Appropriate use of integrated modelling techniques (X = appropriate feature)

	System dynamics	Bayesian networks	Coupled component models	Agent-based models	Expert systems
What is your reason for modelling/type of application?					
Predictive		X	X		X
Forecasting			X		X
Decision-making	X	X	X		X
System understanding	X		X	X	X
Social learning	X		X	X	X
What types of data do you have available/want to use to populate your model?					
Qualitative and quantitative data		X			X
Quantitative data only	X		X	X	
Do you want your model to focus more on a complex description of specific processes in the system or to more broadly cover interactions in your system?					
Depth of specific processes			X		
Breadth of system	X				
Compromise					X
Both		X		X	
Do you want your model to provide explicit information about uncertainty caused by model assumptions?					
Yes		X			X
No	X		X	X	
Are you interested in investigating the interactions between individuals and their impact on the system, or only the aggregated effect of human behaviour?					
Interactions between individuals				X	
Aggregated effects	X	X	X		X

There are some distinct advantages in performing integration methods within an appropriately designed framework and set of tools where the audience and needs are well-identified for their prescription. They can contribute comprehensively to knowledge discovery, brokering, review and management by constituting: a way of investigating and explaining trade-offs; a readily accessible collection of models, methods and visualisation tools; a focus for integration across researchers and stakeholders; a training and education function; an exploratory aid capable of adoption and further development by stakeholders; a permanent summary of the project methods; and a means of making the management analysis transparent.

Finally, integrated assessment is by definition an interdisciplinary exercise requiring good skills. The availability of research training in the techniques of IA can and needs to be accelerated for industry, managers and researchers.

References

Babel MS, Das Gupta A & Nayak DK (2005). A model for optimal allocation of water to competing demands. *Water Resources Management* **19**, 693–712.

Brown DG, Page SE, Riolo R & Rand W (2004). Agent-based and analytical modelling to evaluate the effectiveness of greenbelts. *Environmental Modelling and Software* **19**, 1097–1109.

Castelletti A & Soncini-Sessa R (2006). A procedural approach to strengthening integration and participation in water resource planning. *Environmental Modelling and Software* **21**, 1455–1470.

Feuillette S, Bousquest F & Le Goulven P (2003). SINUSE: a multi-agent model to negotiate water demand management on a free access water table. *Environmental Modelling and Software* **18**, 413–427.

Forsyth R (1984). The expert systems phenomenon. In *Expert systems: principles and case studies* (ed. R Forsyth), pp. 3–8. Chapman & Hall: London.

Gao YC, Liu CM, Jia SF, Xia J, Yu JJ & Liu KP (2006). Integrated assessment of water resources potential in the North China region: developmental trends and challenges. *Water International* **31**, 71–80.

Giupponi C, Jakeman AJ, Karssenberg G & Hare MP (eds) (2006). *Sustainable management of water resources: an integrated approach.* Edward Elgar Publishing, Cheltenham, UK.

Hare M, Letcher RA & Jakeman AJ (2003). Participatory modelling in natural resource management: a comparison of four case studies. *Integrated Assessment* **4**, 62–72.

Harremoës P & Turner RK (2001). Methods for integrated assessment. *Regional Environmental Change* **2**: 57–65.

Haslam SA, Eggins RA & Reynolds KJ (2003). The ASPIRe model: actualising social and personal identity resources to enhance organisational outcomes. *Journal of Occupational and Organisational Psychology* **76**, 83–113.

Hickey JT & Diaz GE (1999). From flow to fish to dollars: an integrated approach to water allocation. *Journal of American Water Resources Association* **35**, 1053–1067.

Jakeman AJ & Letcher RA (2003). Integrated assessment and modelling: features, principles and examples for catchment management. *Environmental Modelling and Software* **18**, 491–501.

Jakeman AJ, Letcher RA, Rojanasoonthon S, Cuddy S & Scott A (eds) (2005). Integrating knowledge for river basin management: progress in Thailand. ACIAR monograph No. 118, Australian Centre for International Agricultural Research: Canberra.

Kidd AL (1987). Knowledge acquisition: an introductory framework. In *Knowledge acquisition for expert systems: a practical handbook* (ed. AL Kidd). Plenum Press: New York.

Land and Water Australia (2006). *Water perspectives.* Outcomes of an expert workshop scoping social and institutional research questions in support of the implementation of the National Water Initiative.

Lanini S, Courtois N, Giraud F, Petit V & Rinaudo JD (2004). Socio-hydrosystem modelling for integrated water-resources management: the Hérault catchment case study, southern France. *Environmental Modelling and Software* **19**, 1011–1019.

Lee N (2006). Bridging the gap between theory and practice in integrated assessment. *Environmental Impact Assessment Review* **26**, 57–78.

Letcher RA & Jakeman AJ (2003). Application of an adaptive method for integrated assessment of water allocation issues in the Namoi river catchment, Australia. *Integrated Assessment* **4**, 73–89.

Letcher RA & Jakeman AJ (in press). Integrated modelling approaches for natural resources management: a review. *Environmental Modelling and Software.*

Letcher RA, Jakeman AJ & Croke BFW (2004). Model development for integrated assessment of water allocation options. *Water Resources Research* **40**, W0552.

Letcher RA, Croke BFW & Jakeman AJ (2007). Integrated assessment modelling for water resource allocation and management: a generalised conceptual framework. *Environmental Modelling and Software* **22**, 733–742.

Martín de Santa Olalla FJ, Domínguez A, Artigao A, Fabeiro C & Ortega JF (2005). Integrated water resources management of the hydrogeological unit 'Eastern Mancha' using Bayesian belief networks. *Agricultural Water Management* **77**, 21–36.

Merritt WS, Croke BFW, Jakeman AJ, Letcher RA & Perez P (2004). A biophysical toolbox for assessment and management of land and water resources in rural catchments in northern Thailand. *Ecological Modelling* **171**, 279–300.

Messner F, Zwirner O & Karkuschke M (2006). Participation in multi-criteria decision support for the resolution of a water allocation problem in the Spree River basin. *Land Use Policy* **23**, 63–75.

Norton JP & Reckhow K (2006). Modelling and monitoring environmental outcomes in adaptive management. In *Proceedings of the IEMSs third biennial meeting: Summit on Environmental Modelling and Software* (eds A Voinov, AJ Jakeman & AE Rizzoli). International Environmental Modelling and Software Society, Burlington, US, July 2006.

Pearl J (1990). *Probabilistic reasoning in intelligent systems.* Morgan Kaufman: San Mateo, California.

Rosegrant MW, Ringler C, McKinney DC, Cai X, Keller A & Donoso G (2000). Integrated economic-hydrologic water modeling at the basin scale: the Maipo river basin. *Agricultural Economics* **24**, 33–46.

Syme GJ, Nancarrow BE & McCreddin JA (1999). Defining the components of fairness in the allocation of water to environmental and human uses. *Journal of Environmental Management* **57**, 51–70.

Ticehurst JL, Rissik D, Letcher RA, Newham LHT & Jakeman AJ (2005). Development of decision support tools to assess the sustainability of coastal lakes. In *Proceedings of MODSIM 05 Conference* (eds A Zerger & R Argent), pp. 2414–2420. 12–15 December, Melbourne.

Ticehurst JL, Newham LTH, Rissik D, Letcher RA & Jakeman AJ (2007). A Bayesian network approach for assessing the sustainability of coastal lakes in New South Wales, Australia. *Environmental Modelling and Software* (corrected proof available online 5 May 2006).

Chapter 9

Delivering the National Water Initiative

The emergence of innovative legal doctrine

Douglas E Fisher

Paragraph 23 of the 2004 Intergovernmental Agreement on a National Water Initiative (NWI) makes clear that its outcome is a system of governance for the sustainable use and development of all water resources in Australia, for whatever purpose they are used (the 'objective') by means of appropriate planning, regulatory and market arrangements (the 'means'). Sustainability embraces economic, social and environmental perspectives. This is a challenge for the legal system because it requires coherent and consistent arrangements across water resources as a whole in ways that are intrinsically enforceable as a matter of law within a federally structured liberal parliamentary democracy.

The arrangements described in the NWI are ambitious and innovative. If the law is to be one of several instruments through which the NWI is implemented, then a number of critical doctrinal changes need to be made to the legal system. This chapter will assess the need for such changes and suggest hypothetical approaches to creating an enforceable legal structure. This involves:

- identifying the critical driving principles
- describing current approaches
- creating hypothetical approaches.

These hypotheses are to some extent novel, take legal doctrine down a slowly evolving path, and are likely to be contentious.

1 The principles supporting a system of governance

It is beyond this chapter's scope to examine in detail the principles and characteristics of an effective legal system. Nevertheless, some brief mention is useful in advance of the hypothetical approaches discussed later.

While ecologically sustainable development drives the management of all natural resources, it is by no means obvious that the detailed arrangements for each individual natural resource should or can be the same, or even similar. What distinguishes the management of water resources from other resources? The distinguishing features include:

- water is essential to all forms of life
- unpredictable availability of water for climatic and meteorological reasons
- major practical difficulties in storing, conveying and making available water for consumptive as well as non-consumptive purposes
- high costs of constructing and maintaining water infrastructure
- a variety of cultural perceptions about water
- major challenges in understanding and managing the hydrological cycle
- close ecological relationships between water, land and the range of other natural resources.

In Australia, the challenges are exacerbated because of Australia's federal system of government, the different systems of management and legal arrangements in each of the states and territories, and the limited ability of the Commonwealth to deal with water management (Fisher 2000). The NWI contemplates a set of related planning, regulatory and market arrangements. The juxtaposition of these three functions – based upon differing legal doctrines – poses particular challenges for an integrated and coherent set of legal arrangements. This is why legal arrangements for the management of resources other than water cannot be easily translated to water.[1] Water, in other words, is unique.

The rights and duties of all agencies and persons involved in an effective legal system must be simple, transparent, stable and predictable. Otherwise those involved will lose confidence in the system and it will be difficult to enforce. However, a system of governance of water resources driven by ecologically sustainable development goes further, involving adherence to five fundamental principles. First, it must be supported by the principle of comprehensive application, meaning that this system of governance applies to all water resources in Australia. This is particularly difficult in a federal system with cross-boundary watercourses and water systems. Second, the principle of comprehensive involvement is designed to bring together all persons and institutions within the community, both public and private sectors, with a view to creating a sense of common or shared responsibility for the sustainable management of water resources. In practice the principle of comprehensive involvement brings together strategists, planners, regulators, water suppliers and consumers, and other interested parties. The third principle – of total systems integration – takes this further, linking strategic, planning and operational decision-making in both procedural and substantive terms. Thus, decisions and activities at the operational level must comply with any enforceable rules created at the planning level.

The effective performance of these strategic, planning and operational functions depends upon public availability of extensive and current information about water resources in general and in specific locations. Hence the fourth principle – information availability. The formulation of strategies and plans that create norms of decision-making and behaviour necessarily requires such information to be available to strategists and planners. Not only that: these processes involve public participation and such information must be available to the public (see Fisher 2006d). Finally, there is the principle of enforceability. However, the enforceability of public rights and duties raises a number of constitutional and institutional as well as legal issues (see Fisher 2000, 2006d).

Incorporating these principles within an effective legal framework extends the scope and function of a legal system in a number of ways. In particular, the principle of enforceability requires a much less descriptive and much more prescriptive set of arrangements. This means, for example, not only conferring power upon all participants but also prescribing how these powers are to be exercised – a prescription of what is to be done as well as how it is to be done. The functions of planning and regulation involve the exercise of powers by the public sector. The exercise of these powers, however, involves the creation of rights and obligations for the private sector. This is true generally, but particularly in the context of the use of market arrangements as an instrument of regulation.

There are a number of public sector agencies endowed with a range of different and potentially conflicting mandates, and the legal system is likely to become more complicated to the extent that it involves accommodation between them. The relevant relationships include those:

- between citizen and citizen
- between citizen and public sector agencies
- between public agencies themselves.

The traditional distinction between public and private legal relationships is breaking down because the governance of water resources is moving towards the public and private sectors acting together to achieve sustainable use and development (Fisher 2006b).

2 Current approaches to managing water resources

The way in which water resources in Australia are managed has changed over time (Fisher 2006a). The establishment of the several colonies from 1788 onwards saw the introduction of the rules of English common law. Essentially, water resources were managed in accordance with rights incidental to the ownership of land, including those associated with the riparian doctrine – a set of private rights and duties. This changed towards the end of the 19th century. For the next century water resources were managed largely in accordance with a system of public management administered by various public agencies exercising largely discretionary powers. This enabled the use and development of water resources mostly for economic and commercial purposes. This changed towards the end of the 20th century, when the legal arrangements were substantially modified to facilitate sustainable use and development of water resources. This has involved an increasingly prescriptive set of rules affecting both the private and public sectors, a return to a system of legal rights and duties.

Each of the legislatures in Australia has enacted legislation for this purpose. Each jurisdiction has adopted its own approach. There is no uniformity of language, structure, procedure or institutions. While each jurisdiction is seeking to achieve sustainability in some form, the means for doing so are different and in many respects incompatible. For example, in relation to strategic issues, NSW has a state water management outcomes plan, SA has a state water plan, WA has an environmental water provision policy and Tasmania has a sustainable development policy.

Each jurisdiction provides for plans of some kind,[2] water rights are structured differently in each jurisdiction, and water rights are created, administered and transferred differently in each jurisdiction. Water rights may assume the form of water allocations, access licences, bulk entitlements, licences to take and use water, resource operations licences and works approvals. The rights and duties associated with each instrument depend upon the terms of the instrument, which in turn depend upon terms of the legislation. The transfer of an instrument in most cases requires approval, and the criteria for approval are not uniform.

The ecological values of water are protected differently. In NSW administrative priority is given to protection of the water source and its dependent ecosystems. In Victoria licences for in-stream use of water are held by the relevant minister. In Queensland the plans contain detailed rules about water sharing. In SA environmental water requirements are determined in accordance with the principles in the legislation. In WA environmental water requirements are imposed as conditions of water licences. Tasmania is guided by its sustainable development policy. In the ACT there are environmental flow guidelines. In the NT the maintenance of the health of aquatic ecosystems is a beneficial use of water.

3 The range of current legal instruments

The capacity of a legal instrument to be enforced depends upon how it is structured. A rule of law that is enforceable sets out with clarity and certainty what may or may not be done, how it must be done or not done and the standards applicable to the decision or the activity. The NWI contemplates a system of governance that comprises a wide range of different instruments. Their capacity to be enforced is variable. The arrangements contemplated by the NWI are in general terms described as strategic, planning, regulatory or market-based. Such an approach is generally not reflected in contemporary statutory structures.

At the strategic level these structures frequently comprise statements of objectives and statements of principles or priorities. At the planning level there is a wide variety of planning instruments: either state-wide, catchment-based or based upon some other criterion such as area, risk or subject matter. Then there are rights in relation to water of various kinds. These are institution-specific, person-specific or location-specific. These rights are exercised within the strategic and planning frameworks already in place. The powers associated with water rights may be exercised in a number of different ways. The right may comprise a share of a water resource, a power to extract water, a power to use water for a certain purpose, or incorporate a power to transfer the right. The rights and duties of a holder of a water right can realistically be determined only through analysis of the strategic, planning, regulatory and market-based arrangements. The relationship between these components of the system is therefore critical. This sense of variety and yet of interdependence emerges from an analysis of how the legislation has been constructed.

3.1 Objectives

First, there are objectives. Section 10(1) of the *Water Act 2000* (Qld) states that its purpose is to advance sustainable management and efficient use of water and other resources by establishing a system for the planning, allocation and use of water. The interpretation of sustainable management in subsection (2) acknowledges the necessary elements of sustainability including the physical, economic and social needs of humans, protection of natural ecosystems, protection of water itself from degradation and recognition of the interests of Indigenous people. The legislation is expected to be administered with community participation and in ways that are integrated with the administration of other natural resources legislation and in accordance with the principles of ecologically sustainable development. Efficient use of water is interpreted in subsection (3) to include demand management measures, water conservation and water quality objectives, and recycling. This statement of purpose is supported by the imposition of a duty upon all entities to perform their statutory functions or exercise their statutory powers in a way that advances this purpose.

The critical elements of this statutory statement of objective are:

- it states comprehensively the various elements of sustainability
- it applies to all water within the jurisdiction
- sustainable management as interpreted is juxtaposed to efficient use as interpreted
- the purpose is achieved through a system for the planning, allocation and use of water
- the purpose is linked to a duty related to its advancement.

3.2 Principles

Second, there are principles. Section 5 of the *Water Management Act 2000* (NSW) states the water management principles of the Act. Subsection (3) states the principles in relation to water sharing. Section 9(1)(b) imposes a duty on all persons exercising functions under the Act to give priority to these principles for water sharing in the order in which they are set out in section 5(3). In effect, the priority is this:

- first is the protection of the water source and its dependent ecosystems
- second is the protection of basic landholder rights
- third is the sharing or extraction of water in accordance with any other right.

For this purpose basic landholder rights are domestic and stock rights, harvestable rights and native title rights.

The significance of this approach is fourfold. First, it is constructed as a duty imposed on all persons exercising functions under the Act. Second, the duty is to give priority to the relevant principles in the order set out in the Act. Third, the protection of the water source and dependent ecosystems is given first priority. Fourth, these statutory provisions have been constructed so that the courts are prepared to recognise a justiciable issue for their determination. Significantly, therefore, section 3's general statement of the Act's objects about the sustainable and integrated management of the state's water sources is applied subject to and in accordance with the duty imposed by the juxtaposition of section 5(3) and section 9(1)(b).

3.3 Plans

The water resources of SA are managed in accordance with the *Natural Resources Management Act 2004* (SA). Water resources are a component of natural resources for this purpose. The Act provides specifically for the formulation of a range of plans:

- the State Natural Resources Management Plan
- regional natural resource management plans
- water allocation plans.

Each of these plans is linked to the others and to achievement of the objects of the Act in general. Section 8 places a duty on all institutions and persons involved in the administration of or performing functions under the Act to have regard to the objects of the Act and to seek to further them. Significantly, this is complemented by a more specific duty placed upon all agencies and instrumentalities of the Crown to endeavour, as far as practicable, to act consistently with the State Natural Resources Management Plan and all other relevant natural resources management plans under the Act.

The state plan deals with water as a natural resource. A regional plan must explain how achieving the goals in the plan will assist to achieve the objects of the Act and, more importantly, the objects of the *River Murray Act 2003* (SA) and the objectives for a healthy Murray River under that Act if the plan applies within the Murray-Darling Basin (section 75(3)(c)(ii)). In addition, a water allocation plan must be prepared for each of the prescribed water resources in the region of the regional natural resources management board. Section 76(4) prescribes in detail what must be included in a water allocation plan. The information upon which a water allocation plan is based is contained in the regional plan. It is, therefore, the intrinsic function of a water allocation plan to provide for the allocation and use of water on the basis of that information so that an equitable balance is achieved between environmental, social and economic needs for the water, and the sustainable rate of use of the water (section 76(4)(b)).

The significance of this approach is fourfold. First, there is the need for consistency between the water allocation plan, the regional plan and the state plan. Second, there is the close relationship between each of these plans and the objects of the Act. Third, there is the generic statement of the objects of the Act and the correlative duties placed upon those administering or performing functions under the Act. Fourth, the criteria for determining water allocations and water uses are relatively general: an equitable balance between the various needs for the water and the sustainable rate of use of the water. The standard is thus one of equity and not of priority.

3.4 Water rights in general

The way in which water rights are structured is critical to their tradeability, but also their enforceability. A water right has traditionally been structured either as a right to take or extract and use the water (frequently with a right to construct the necessary infrastructure) or a right to receive water from an institution or a person entitled to extract and use it for the relevant purpose. The structure of water rights is, however, changing to include a right in the nature of an interest in the water resource, to which there may be linked a more traditional right to take, extract, use or receive. Interests in water and in water resources generally and rights associated with them are thus increasingly being doctrinally separated or disconnected to enable water resources to be managed in accordance with a range of instruments, including those associated with a market. The evolution of differently structured interests in water has become quite a subtle process.

3.5 Rights to take water

Under the *Rights in Water and Irrigation Act 1914* (WA), a licence is required to take water from any watercourse, wetland or underground water source to which the section applies (section 5C(1)(d)). According to section 55 of the *Water Management Act 1999* (Tas.), a licence authorises its holder to take water in accordance with the licence and is subject to any relevant water management plan and any conditions specified in the licence.

In South Australia a licence is endorsed either with a water (taking) allocation or a water (holding) allocation. Under the *Natural Resources Management Act 2004* (SA), a licence endorsed with a taking allocation authorises its holder to take water from the prescribed watercourse, lake or well or to take surface water from the surface water prescribed area specified in the licence (section 146(1)). In the case of a licence endorsed with a holding allocation, the holder of the licence is authorised not to take the water but to request the minister to convert the allocation to a taking allocation.

In the NT, under the *Water Act 1992* a licence is required to take or use water flowing or contained in a waterway (section 44). Similarly, in the ACT the *Water Resources Act 1998* specifies that a licence is required to take water (section 33(1)). However, a licence to take water must not be granted unless a water allocation or an interstate water allocation exists on which to base the taking of water from the place to which the application for a licence relates (section 35(8)(a)).

3.6 Rights to receive water

A right to receive water is structured differently. A right – or at least an expectation – to receive water is conferred mostly in relation to a supply of water for household purposes in urban communities, or in relation to a supply of water for irrigation in rural communities. The interest of an irrigator in Victoria under these arrangements is set out in section 222(1)(b) of the *Water Act 1989*. The institution authorised to supply water in an irrigation district must make available for supply to the owner or occupier of each holding in the irrigation district the amount of water for irrigation specified in the register in relation to that holding. When the arrangements enacted in 2005 are in force (Victoria 2005), the duty will become a duty to provide the service of delivering water to the owner or occupier of each serviced property in the irrigation district for the purposes of irrigation at the volumes and for the periods determined in accordance with the arrangements in this part of the Act. In each case – under existing or amended arrangements – the supplier has the power to reduce, restrict or discontinue the supply of water in the circumstances described in the Act. The statutory duty to supply water for irrigation purposes is thus conditional, in practical terms.

3.7 Rights in relation to water

Water rights in NSW may be described as rights in relation to water rather than water rights themselves, created through four instruments of regulation under the *Water Management Act 2000* (NSW). There are access licences which are divided into two components, water use approvals and water management work approvals. The holder of an access licence is entitled to specified shares in the available water within a specified water management area or from a specified water source – the share component (section 56(1)(a)). An access licence also entitles the holder to take water at specified times, at specified rates or in specified circumstances, or in any combination, and in specified areas or from specified locations – the extraction component (section 56(1)(b)). A water use approval confers a right to use water for a particular purpose at a particular location (section 89(1)). A water supply work approval authorises its holder to construct and use a specified work at a specified location (section 90(2)). The significance of this approach is the creation of the share component independently of the right to take water, to use water and to construct and use works related to water.

A similar set of concepts will emerge in Victoria when the arrangements enacted in 2005 become effective. The regulatory instruments will be a water share, a water allocation, and either a water use licence or a water use registration. A water share will authorise taking water under the water allocation for the share during the water season for which the allocation is allocated or, with the approval of the minister, in a subsequent season (section 33F(2)). A water allocation will be determined by the minister for a declared water system for each water share for which the system is the associated system (section 33AC(1)). It comprises the amounts of water that are to be allocated to that share from the water determined to be available for allocation to water shares in the seasonal determination. A water use licence will authorise the use of water to be used for irrigation on land provided the water is authorised to be taken in accordance with a water share and a water allocation. A water use registration will similarly authorise the use of water on land for purposes other than irrigation.

3.8 Rights to transfer water rights

Water rights are structured to perform particular functions within the system of governance. Interests and rights in relation to water are being separated for a number of purposes, including the facilitation of market trading. Water rights have traditionally been transferable but usually only with the approval of a public sector agency granted in accordance with prescribed criteria. One of the outcomes contemplated by the NWI is the facilitation of trading in water rights. However, trading in water rights requires more than their transferability. It has been suggested that an effective set of market arrangements requires:

- individual ownership of the asset to be traded
- full and comprehensive information about the asset and the rights and duties associated with ownership of the asset
- a system of competitiveness (Fisher 2006c),

A water right therefore needs to be structured to reflect these characteristics and the procedures for trading arranged accordingly.

3.9 Environmental water entitlements

The ecological health and values of water resources are conserved in a number of different ways (Gardner 2006): in plans, as conditions in licences or as environmental water entitlements. The *Water Act 2000* (Qld) requires a water resource plan to include the ecological outcomes for the sustainable management of the water (section 46(1)(e)). A resource operations plan must state how the water to which the plan applies will be sustainably managed (section 98(1)(e)). This may be done by including environmental management rules within the plan. In practice, these rules include strategies for meeting environmental flow requirements. In WA, the *Rights in Water and Irrigation Act 1914* (WA) states that the terms or conditions of a licence may relate to the use, management, protection and enhancement of any water resource and its ecosystem or the environment in which the water resource is situated (Schedule 1, appendix).

The *Water Act 1989* (Vic.) indicates that the objective of setting aside water for the environment is to preserve the environmental values and health of water ecosystems, including their biodiversity, ecological functioning and quality of water and the other uses that depend on environmental condition (section 4B(1)). The environmental water reserve is the water set aside for this purpose in accordance with section 4A. The ways in which this is achieved include:

- as an environmental entitlement
- as a condition in a licence
- in accordance with a management plan.

Section 52(1) provides for an environmental entitlement in the form of a licence for an in-stream use of water. When the arrangements enacted in 2005 become effective, in-stream use licences will be replaced by allocations of environmental entitlement. An entitlement may be allocated to the Environment Minister on behalf of the Crown. The purpose of such an allocation is to maintain the environmental water reserve in accordance with the environmental water reserve objective or generally improving the environmental values and health of water ecosystems (section 48B(1)).

Conclusion

These arrangements disclose a set of mosaics or patterns that are structurally linked in different ways, but each lead to the sustainable use and development of water resources in one form or another. While there is no thematic or doctrinal consistency – let alone uniformity – each set of arrangements strives to achieve its objectives in its own way. This is particularly true in relation to the way water rights are structured – a matter of particular significance in the context of arrangements for trading in water rights.

4 A hypothetical set of structural arrangements

If the outcomes of the NWI are to be achieved, what might a set of generic arrangements look like? A more detailed set of proposals is discussed in Fisher (2006d). It is important that these arrangements reflect the five principles already discussed:

- the principle of comprehensive application
- the principle of comprehensive involvement
- the principle of total systems integration
- the principle of information availability
- the principle of enforceability.

These principles need to be incorporated within the structural framework described by the NWI, namely, a set of planning, regulatory and market arrangements. Each of these three functions must therefore be included within the overall structure created for the achievement of the sustainable use and development of all water resources in Australia. Let us now consider the range of hypothetical structures that may lead to an enforceable legal regime for sustainable water resources governance.

4.1 Strategies and outcomes

First of all, the legislation itself will state the overall strategies and outcomes for the governance of water resources within the jurisdiction. Most importantly, it will state all of the strategies and outcomes. However, the nature of sustainability is such that the system will be directed at the achievement of one and only one outcome, namely, the sustainable use and development of water resources. This outcome almost necessarily involves a balancing function among all the competing principles and criteria that relate to any decision-making process. While sustainability is itself an outcome, by its very nature it does not contain any priorities. One of the functions of the legislation may well be to state priorities within the overall concept of sustainability. In any event, the statement of strategies and outcomes in the legislation is likely to be complemented by what may be described as paralegal rules, that is, norms of decision-making such as statements of principle, statements of guidelines, statements of criteria, statements of value and more specifically statements of ways and means by which these outcomes may or perhaps must be achieved. This involves a subtle combination of legal rules and paralegal rules within an overall prescriptive structure (see Fisher 2005, 2006a).

4.2 Plans

Next, the legislation will provide for the formulation of plans at several levels – state or territory-wide plans, catchment plans or local plans (see also Chapter 5, this volume). Much depends upon hydrogeological and hydrological considerations. The legislation will provide for how these plans are made. Unquestionably this will involve public participation at all levels. Legislation will provide for the substance of these plans. How this is done is important. There are a number of possibilities. The legislation may state that the matter is for consideration, or state what matters must be addressed. It may state the specific outcomes of the planning regime. Just as important, these plans must be formulated substantively in accordance with the strategies and outcomes stated by the legislation. In other words, there must be substantive consistency and coherence between the high-level statements of strategy and outcome and the implementation of these strategies and outcomes at the level of planning. It is likely that the legislation will require specific uses of the water resources to be addressed in the plans, for example, consumptive, environmental or cultural uses.

While the plans are essentially concerned with allocating the water resource to various uses, the relationship between plans and instruments of regulation cannot be ignored. The plans, in other words, will contain rules according to which the function of regulation is performed. The rules contained in the plans will enable those involved to know their rights and duties with as much certainty as possible. For example, whether an application for a water right or a right in relation to water will be granted or not, on what terms and conditions, whether it is able to be transferred and on what conditions, and whether the right is likely to be interfered with and on what conditions and with what consequences.

4.3 Regulation

Then there is the function of regulation. Access to water resources is granted in accordance with the strategies and outcomes stated in the legislation and in the enforceable rules contained in the plans. Rights granted in this way are specific to the holder of those rights, the water in question, and the use and location of the water. Rights of access in this sense and the operational rights to which they are linked arise in specific sets of circumstances but in accordance with the strategic arrangements in the legislation and plans. It is these rights that are the subject of market arrangements. For a market to be effective as an instrument of governance, the nature and subject matter of these rights must be clear, certain and predictable. Whether they have these characteristics is a reflection not only of their own nature but also of the operational powers and capacities associated with these rights. These derive from other elements of the legal framework including the strategies and outcomes in the legislation and more particular rules in the plan. A system of governance must therefore reflect the relationship between all the elements it comprises. Some are the responsibility of the public sector, while others are the function of the private sector. But in all cases the ultimate responsibility lies with the public and private sectors operating together in accordance with this linked and coherent set of arrangements that incorporates these legal and paralegal rules.

5 A statutory charter to implement the NWI

5.1 Policy issues

The formulation of a statutory charter for the governance of water resources is a considerable challenge. We will use the terminology in the NWI. Before a set of generic arrangements can be identified, a number of threshold questions need to be addressed at the level of policy (Fisher 2006d):

- the range and nature of water access entitlements that may be created, for example consumptive, environmental or cultural
- which water access entitlements are transferable and which are not
- who is entitled to hold a water access entitlement, for example any person, an accredited person or a person authorised by a public agency
- how a water access entitlement is created, for example directly by a water plan, in accordance with a water plan or by administrative discretion
- how a water access entitlement is transferred, for example in the exercise of an unfettered discretion by the holder, in accordance with rules in a water plan or with the approval of a public agency.

Similar questions arise in relation to water allocations and water use approvals.

5.2 A fundamental duty

Just as important is the relationship between water access entitlements, water allocations and water use approvals and the relationship between these and water plans and the overall strategy contained in the legislation. There emerges, therefore, a set of interlinked individual rights and duties which are themselves part of a coherent strategic structure. This is likely to be achieved only by the creation of a fundamental duty structured so that it is enforceable. This duty needs to be imposed upon all persons and institutions involved in the governance of water resources, and this specifically includes the relevant public sector agencies and private sector participants. A duty along these lines has been suggested:

> It is the duty of all persons and institutions (including the state or the territory and its administrative, executive and judicial agencies), exercising powers, performing functions or fulfilling obligations under this Act, any plans made under this Act and any water management instruments granted under this Act to achieve the objective of this Act, to observe the principles stated by this Act, to comply with the rules contained in or provided for by this Act, to comply with the plans made under this Act and to comply with the water management instruments granted under this Act or, as the case may be, to ensure the achievement of the objective, the observance of the principles and the compliance with the rules, the plans and the water management instruments (Fisher 2006d).

This is an ambitious statement of legal doctrine, but one that is necessary if a comprehensive, coherent and enforceable system of governance of water resources is to be achieved in accordance with the NWI.

5.3 Duties in relation to information

The creation of strategies and policies, the formulation of plans and the making of decisions in particular sets of circumstances depend on the quality of the information upon which they are based for their effectiveness. Information in support of these functions is located in a range of different sources. Public sector agencies at federal, state and local level have access to a wide range of information about water resources, as do private sector participants in the water industry. Individuals within the community who extract or receive water have information about the source of supply, the purpose for which it is used and how much is used. Just as important is information about the rights and duties of all involved in the management and use of water resources. This is particularly true in the context of trading water rights in a market context.

Comprehensive and up-to-date information is therefore essential to support the planning, regulatory and market arrangements contemplated by the NWI.

All institutions and persons involved need access to this information to ensure effective planning, regulation and trading. Some information may be sensitive, confidential or difficult to obtain. Hence the need for a duty on all relevant participants to make generally available the information that is available to them individually. The existence of such a duty is essential to the existence of a correlative right of everyone involved to have this information made available. The information which is the subject matter of these duties is of different kinds:

- information supporting the strategies and plans formulated by public sector agencies
- information about the practical day-to-day operation of the system on the basis of actual use of water resources
- information about individual water rights and rights in relation to water
- information about dealings with water rights and rights in relation to water.

The duty to provide this information must be not only comprehensive but ongoing. There are various ways in which a range of appropriate duties can be constructed. This involves the creation of a public water resources information and accounting system, but it needs to be a system that makes information available to the public in a meaningful and transparent way. This becomes ultimately the responsibility of the public sector, upon which a number of more specific duties need to be imposed. These include:

- a duty to collect and collate the information
- a duty to analyse the information
- a duty to present the information according to acceptable standards
- a duty to do so in a form and in a way that members of the public can understand.

Summary

If such a set of statutory arrangements is to be enforced as a matter of law then the rights and duties that lie at the foundation of the system must be structured accordingly, issues of a justiciable nature must be created, procedures for enforcement must be included in the legislation and a range of sanctions must be imposed for a non-compliance. In summary, a generic statutory structure may be divided into ten parts:

- foundational elements
- a system of rights and duties
- information availability
- water plans
- water access entitlements
- water allocations
- water use approvals
- urban water
- transactions
- enforcement.

6 Implementation challenges

6.1 A fundamental duty

The last few years have seen a dramatic change in the way the law permits or requires water resources to be managed. A relatively uncontrolled discretionary system has been increasingly replaced with one incorporating enforceable rules of a procedural nature coupled with statements of outcome or objective whose status is much more equivocal. Sustainability is by its nature an outcome. If this is to be achieved as a matter of law, then the objective of sustainable use and development needs to be incorporated within a matrix of enforceable rights and duties. In some cases legal arrangements are moving in this direction. New doctrinal developments of this kind present opportunities as well as challenges for the judicial system. It is possible to construct a statutory duty for this purpose along the lines suggested. Whether this is acceptable as a matter of policy is another matter.

6.2 Statements of priority

The essence of sustainability is to seek an outcome that balances all the relevant considerations. In this sense, the outcome in an individual case depends entirely on its own circumstances. In other words, while the law may direct the achievement of sustainability, it currently does little more than indicate all the factors to be considered in achieving this. The law generally does not state clearly and categorically the strategic preferences and priorities. It should do so. Consider an example.

The NWI seeks to prevent overallocations in future and to reduce overallocations granted in the past. This is a sensitive issue to manage, particularly in the context of overallocations for non-domestic consumptive use. It should be stated in the legislation. It might be structured as an enforceable statement of priority or as a duty imposed upon a minister or upon the administration of the legislation generally. In this event, the priority would be created as a matter of law by the legislature and decisions made by the minister or other member of the executive in compliance with the duty created by the legislation. It is a matter of policy whether the legislature would wish to take the responsibility for creating priorities of this kind.

6.3 Stability and adaptability

It has been suggested that the principle of comprehensive involvement brings a sense of common or shared responsibility for the sustainable management of water resources. This is particularly important in the context of the juxtaposition of arrangements for the consumptive and non-consumptive uses of water. Consumptive uses of water require security of supply and stability of legal arrangements, particularly in the context of market arrangements. On the other hand, the conservation of the ecological values of water and the protection of the environment generally require adherence to the principle of adaptive management. In many respects, security and stability are the antithesis of adaptability. Stability implies non-intervention by government while adaptability implies potential or actual intervention. It is a challenge to create an appropriate balance between these potentially conflicting approaches in an integrated system of governance. In legal terms, security and stability involve protection of legal rights whereas adaptability involves interference with existing legal rights. So the ways in which existing legal rights may be amended or taken away need to be identified with clarity and precision so that the circumstances of intervention can be predicted without doubt or ambiguity. This involves a regime of interference based upon enforceable rules rather than upon the exercise of discretions. There is also the factor of compensation for interference with or removal of existing legal rights. If issues such as overallocation are to addressed effectively, legal arrangements need to be structured to deal with them.

6.4 Information

The principle of information availability is critical to an integrated system of governance directed towards common or shared responsibility for sustainable management of water. In practice this requires the imposition of a wide range of duties upon both the public and private sectors, but in different ways. The legal arrangements make provision for the availability of information in a number of limited ways. For the most part these include providing information to public agencies about holdings and transactions in relation to water rights. A much more detailed and comprehensive set of obligations is necessary in both the private and public sectors to bring about a public water resource information and accounting system, such as that discussed above. While such an approach is no doubt a sensitive matter of policy, it would be particularly difficult to incorporate an effective system within a legislative structure.

6.5 Uniformity or distinctiveness

Each state and territory in Australia has its own regime for the governance of water resources. There is limited or in some cases no acknowledgment of the interjurisdictional complexities of water resources management. The Commonwealth has no direct involvement in managing water resources and its capacity to do so indirectly is limited. Nor has there emerged a set of common law rules governing the management of transboundary water resources. The arrangements in place in relation to the Murray-Darling Basin and the Lake Eyre Basin are essentially political rather than legal arrangements. It is no surprise therefore that there is no uniformity and relatively little consistency in the legal arrangements in the six states and the two territories.

Each jurisdiction is moving in the same direction, namely, sustainable use and development of water resources. But the strategic, planning, regulatory and market-based arrangements are different, sometimes substantially. For example, in some cases legislation applies to all waters comprehensively within the jurisdiction while in others the legislation applies in prescribed contexts. Similarly, there is no uniformity about the rules of access to the judicial system for settling disputes or testing the legitimacy of government decisions. Even more importantly, water rights and rights in relation to water are structured in different ways that affect the nature of the asset thereby created, and this clearly affects the economic and commercial value of the asset. This is again particularly significant for interjurisdictional trading in these rights.

6.6 Integration in a federal system

The achievement of the sustainable use and development of water resources depends to some extent upon the legal arrangements designed to bring it about. Sustainability as a concept is the outcome of a system of governance which must essentially be integrated in the ways described. Integration in a legal sense means one simple yet coherent system of legal arrangements designed to achieve the same purpose. The instruments or mechanisms are strategic, planning, regulatory or market-based – each performs a different and unique function within the overall system but must be linked to the others in a completely integrated and coherent way. This is difficult in any event, but it is particularly difficult to achieve in a federal system.

Conclusion

The principal challenge for the implementation of the NWI is the incorporation of ecologically sustainable development as an enforceable duty within a relevantly structured statutory charter for the governance of water resources. This depends upon three critical factors:

- integration of strategic, planning, regulatory and market-based functions
- creation of sustainability as a justiciable issue
- clear definition of the legal rights and duties of all persons and institutions, public and private, performing functions.

There are also a number of more specific challenges:

- creation of statutory priorities in the allocation of water resources to particular uses
- creation of a range of general and specific duties in relation to the provision and availability of information
- creation of detailed arrangements for trading in water rights

- creation of specific and enforceable arrangements in relation to environmental water entitlements
- creation of enforceable rights and interests in relation to the cultural perspectives of water, particularly Indigenous perspectives.

The NWI is an ambitious statement of policy. Its implementation through the legal system requires the development of innovative legal doctrine. This chapter suggests how this may be done. Whether the approaches hypothesised here are feasible and acceptable from other perspectives is a different question altogether.

References

Fisher DE (2000). *Water Law* pp. 37–46. LBC Information Services, Sydney.

Fisher DE (2005). Legal and paralegal rules for biodiversity conservation: a sequence of conceptual, linguistic and legal challenges (Part 1). *Environmental Law and Management* **17**, 243–253.

Fisher DE (2006a). Legal and paralegal rules for biodiversity conservation: a sequence of conceptual, linguistic and legal challenges (Part 2). *Environmental Law and Management* **18**, 35–50.

Fisher DE (2006b). Water law and policy in Australia: an overview. *Environmental Law Reporter* **36**, 10264–10276.

Fisher DE (2006c). Markets, water rights and sustainable development. *Environmental and Planning Law Journal* **23**, 100–112.

Fisher DE (2006d). *A generic set of arrangements for managing interests in water*. Queensland University of Technology: Brisbane.

Fisher DE (2006e). Water resources governance and the law. *Australasian Journal of Natural Resources Law and Policy* **11**, 1–41.

Gardner A (2006). Environmental water allocations in Australia. *Environmental Law Reporter* **36**, 10319.

Victoria 2005. *Water (Resource Management Act) 2005,* being Act No.99/2005 of the Parliament of Victoria.

Endnotes

1 The fisheries sector can offer important lessons for the establishment of effective water markets. See Chapter 6, this volume.

2 In NSW there are management plans. In Victoria there are management plans for water supply protection areas, and non-statutory stream flow management plans. In Queensland there are water resource plans and resource operations plans. In SA there are catchment water management plans and water allocation plans. In WA there are regional, subregional and local area management plans. In Tasmania there are water management plans. In the ACT there are management plans and in the NT there are water allocation plans and water control districts.

Chapter 10

Delivering the National Water Initiative

Institutional roles, responsibilities and capacities

Daniel Connell, Lisa Robins and Stephen Dovers

1 Introduction

The National Water Initiative (NWI) proposes a revolution in Australian water management, a quantum leap in the magnitude and complexity of water management and its related institutions. The necessary institutional reform needs to be understood in the context of this change. This chapter examines the new system of rules the NWI proposes for water management, and argues that the adoption of the NWI has involved a paradigm shift in the underlying logic of water management from community development to one based on property rights.It has particular implications for information needs, skills development, resourcing and regional capacities to support delivery (Connell 2007).

1.1 How is the context changing?

For over a century Australian water management was run by public officials applying an administrative approach to the distribution of heavily subsidised water, and governments and their officials led the way in promoting increased water use. As part of the nation-building project that began in the mid 19th century, the aim was to use water to create new rural communities. Governments actively sought people who would use the water made available by the publicly funded storages and distribution systems. Data about the volumes of water available, where it came from and where it was going, were important for managers, but except in times of drought there was little concern about over-extraction and compliance issues or the need to balance competing priorities.

In each region the managers of such schemes had considerable autonomy in the way they expanded the use of irrigation water. The result was the ad hoc development of many poorly documented entitlement and distribution systems that reflected biophysical variations between regions and the idiosyncrasies of local communities and water managers. For a number of decades this was less of a problem than it might seem. The expansion was occurring during a 50-year period that was significantly wetter than the first half of the 20th century. Extractions were moderate and there was minimal water trading, which meant that inconsistencies between regions were not a problem.

Today, the context for managing water has altered dramatically. Robins (2007b, in press) describes the major paradigm shifts in natural resources management in Australia more generally, including the influences of the landcare movement, integrated catchment management, sustainable development, neoliberalism and regionalisation. Over more than a decade, Australian governments, both Liberal and Labor at federal and state levels, have undergone a philosophical transition from 'nation- and community-building' to a much tighter focus on efficiency and economic growth. At the same time there has been an emerging consensus, in Australia and internationally, that many hydrological systems are now in a state of ongoing environmental decline. Australian concerns were reflected in the imposition of the cap on extractions in the Murray-Darling Basin in the mid 1990s (Murray-Darling Basin Ministerial Council 1995), the 1994 Council of Australian Governments rural water reform program (COAG 1994) and most recently in the NWI. There has also been an expansion in the number of stakeholders seeking to influence water policy and management. These industry, community and interest groups are demanding detailed information and greater input to decision-making processes and, increasingly, organising themselves into powerful advocacy groups able and willing to contest decisions through political and legal action.

1.2 Implications of a property rights regime

The changes that will result from implementation of the NWI will be profound. Australian governments are using their control of water management to create a rights-based regime of tightly defined entitlements for recognised stakeholders ranging from irrigators and towns to the environment itself. The new system has very different requirements regarding the quality, range and volume of information that it needs to function compared with the old administrative system. Vague definitions and imperfect information are now a much greater threat to the operation of the water management system than in the past (Chapter 9, this volume). Then, even though available data were often seriously inadequate and the rights and responsibilities of the various stakeholders involved in a given issue only vaguely defined, water administrators could make decisions for better or worse, based on their considered assessment of the issues. That their aims were usually similar to those of the communities within which they worked reduced the potential for conflict. However, with a growing number of interest groups involved in water management such a consensus is no longer possible.

The NWI's envisaged rights-based system is meant to be largely self-regulating, with reduced scope for discretionary decision-making by administrative officials. When disputes become intractable they are likely to end up in the courts, rather than being solved by administrators or politicians as in the past. This possibility of legal recourse will shape the way rights and responsibilities have to be defined and the nature of the data needed for their administration. Previously accepted standards of data and definitions of rights and responsibilities will be inadequate. As the history of land and house trading indicates, creating and resourcing institutional processes adequate to meet the new requirements will be a complex task.

This upgrading of standards, creating new information systems and reshaping the foundations of water management is being undertaken at a time of considerable confusion. While the federal and state levels of government are heavily involved in water management, the relationship between them is in a state of flux. The 1994 COAG water reform program, a landmark multi-government agreement on water management, has been superseded by the NWI which is being administered by the National Water Commission. The 1994 rural water reform process used national competition payments to encourage compliance by the states. A replacement compliance system will need to be worked out within the new context created by the federal government's $10 billion rural water plan.

1.3 Regional delivery

Contemporary water management is characterised by the involvement of many institutions often with vaguely defined and overlapping roles. The independent centres of sovereign power provided by the Commonwealth and state jurisdictions create focal points around which contending interest groups arrange themselves, moving from one to the other as their members make strategic decisions about alliances and how best to promote their interests. In practice, decisions are not made through a top-down process but are the product of complex interactions in which participants have varying degrees of influence but no single voice is dominant. For any particular region with its own characteristics, the most obvious point where all these contending forces and influences can be integrated is at the regional level rather than national or state. Many spheres of governance have an interest, but none have the same need to coordinate as the people who will have to live with the immediate consequences.

In the NWI, the meeting-point for all the competing tensions involved in water management is the water plans that have to be prepared for each hydrological system. Water planning is the central organising device for the NWI but progress with implementation has been slow. The October 2006 National Water Commission newsletter reported the findings of the Australian Water Resources 2005 audit. At that time, only 18% of surface water management areas and 33% of groundwater management units had a draft or final management plan. A further indication of the work that needs to be done was provided recently by Peter Cullen, a National Water Commissioner. Speaking at the Brisbane Riversymposium in September 2006, he stated that he doubted whether 'we yet have any water plans that are NWI compliant'.

Ultimately, the institutional agenda explicit and implicit in the NWI can be reduced to the need to create an institutional system from the national to the regional level that can develop and implement comprehensive and effective water plans. It is through their preparation that the difficult issues involved in balancing environmental sustainability and economic production are to be resolved. An example of the complexity to be managed in future is the process agreed for bringing activities in dryland parts of catchments into the new water entitlement system (NWI paras 55–57). At a time when the sharing of water resources is becoming increasingly contested and research has shown that some dryland (non-irrigated) land uses, such as lucerne pasture and farm forests, have significant impacts on in-flows (van Dijk et al. 2006; Dawes et al. 2004), it is understandable that governments would want to extend their regulatory regimes to manage these factors. Left unchecked, they could have a substantial negative impact on the quantity and quality of water supplied to towns, irrigators and other users and on the capacity to protect the environment.

Under the NWI, as soon as extractions in a region with a water plan exceed levels that are environmentally sustainable, 'or the system is approaching full allocation', new production activities that will cause significant further reductions in in-flows to surface or groundwater systems will need to purchase a water access entitlement. (If overallocation is not a problem such a purchase will not be necessary.) The process of defining the sustainable extraction level is difficult enough, but how close to full allocation is the point when purchase of water entitlements becomes necessary? The formula also refers to allocation rather than the actual rate of extraction; historically, there have often been substantial differences between the two. Also required will be a better understanding of hydrological systems than is available in most catchments, or it will not be possible to define the threshold beyond which purchase of an entitlement is necessary. Under most established water management systems, monitoring was focused on in-stream flows and the distribution points to and from irrigated properties. The monitoring systems that will be developed as part of the package to be financed by the $480 million assigned to the Bureau of Meteorology under the new Commonwealth rural water reform plan will need to take into account activities in dryland areas of catchments in ways that must be much more comprehensive than before. In the many catchments experiencing development of the type dealt with by paragraphs 55–57 of the NWI (plantation forestry etc.), it will be necessary to record water movement over and through the landscape in ways that were previously only documented under research conditions (van Dijk et al. 2004).

Distinguishing between overallocated catchments and non-overallocated catchments is understandable as a political strategy designed to reduce opposition to the introduction of the new system, but it may create a legal quagmire. How will disputes about such issues be managed under the new system, with its greater clarity of rights and better avenues for appeal not only for those directly involved but also for third parties who previously had minimal opportunity to object? The fact that the conflict will be over rain falling on the area in question rather than water diverted through pipes and channels will add to the tension. Until recent changes regarding the regulation of overland flows in NSW and Victoria, rainfall has not been estimated and made subject to detailed external management and potential water charges. Significant cultural change within rural communities will be required to win acceptance. It is also likely that this new system will create strong economic incentives for the managers of activities such as new forestry plantations to take legal action to resist the transition from the below-allocation regime where rain water is free and apparently unlimited, to the near and above-allocation situation. The simplest way to oppose such a decision will be to question the science and processes used to calculate the threshold. The new system of water management will need to provide appropriate data and scientific understanding so that decisions about conflicts will be well-informed. In addition, it will need robust institutional frameworks that will promote speedy resolution and minimise the potential for delays.

2 Current knowledge

As Fisher (Chapter 9, this volume) makes clear, the NWI is, in essence, an ecologically sustainable development implementation task. That task involves integration of environmental, social and economic policy over the long term, establishing strong public participation processes, generation of high-quality knowledge from diverse sources, constructing innovative and coordinated policy instrument packages, and protecting ecological systems in a precautionary fashion. These tasks are set out, in a particular context, by the NWI. The institutional settings within which this is being attempted – across water and many other sectors – in the past were constructed in response to different imperatives. *Prima facie*, they are likely to be inadequate. The challenge of institutional change for sustainability has been articulated in the international policy arena now for two decades (WCED 1987: 9): 'The real world of interlocked economic and ecological systems will not change; the policies and institutions concerned must'. Yet fundamental institutional change has been rare – change has largely been at the level of organisational revision and experimentation with new policy programs and instruments (Lafferty 2004; Lenschow 2002; Connor & Dovers 2004; Young 2002).

From the partial experiments analysed to date, and from the literature, we can identify core challenges in institutional design for sustainability, all of which are relevant to water and the NWI, and which inform the ensuing discussion in this chapter.

- First, much discussion focuses on organisational and policy change under the rubric of 'institutional reform', with far less attention paid to the importance of more fundamental elements of the institutional system – law, property rights, constitutional rules etc. These rules of the games are more enduring and less quickly changed, but they determine the behaviour of the more easily altered organisations which manifest them and the policy processes and programs which these organisations design and implement.
- Second, individual institutions and individual cases of organisational reform have been the main consideration, with less attention paid to the totality of institutional reform and issues of coordination. In complex policy domains, such as integrated water management, it is helpful to consider institutional systems – interacting systems of policy actors, statutory frameworks, organisations and policy processes and

programs – which are more than the sum of their parts (negatively when not functioning in a coordinated fashion, positively when operating synergistically).

- Third, the centrality and difficulty of policy integration in political and administrative systems that have evolved multiple, specialised and fragmented parts (portfolios, sectors, agencies). The search for integration in disintegrated administrative systems is a prime challenge, especially while preserving the benefits of specialist capacities and knowledge.
- Fourth, the importance of clearly characterising new institutional and administrative functions demanded by sustainability, then assessing existing institutional forms in terms of their capacity to undertake those functions.
- Fifth, the allocation of legal, policy and management responsibilities in multi-tiered policy systems – especially federal systems – remains contested. It has been most explicitly discussed in Europe under the heading of 'subsidiarity'. This issue is complicated when new, administratively and legally weak scales of governance emerge, as in the case of regional organisations in Australian natural resource management.
- Sixth, the magnitude and time-scale of the nature of institutional reform to address sustainability, which is increasingly being understood as a reiterative generational task, unlikely to advance rapidly but requiring a long-term agenda of improving understanding, assessing progress and incremental yet purposeful institutional redesign.
- Lastly, the close and mutually dependent links between normative change and institutional change. Reform of formal institutions that outpaces the understanding and expectations of the population, or vice versa, is likely to be unstable and uncertain of continued progress.

Putting in place better institutional and organisational structures, however difficult, is only part of the challenge. This institutional 'hardware' will not perform as expected in the absence of the institutional 'software' (Dryzek 1996) required to run it – the human intelligence, knowledge, skills and organisational culture demanded by new functions.

It does not take long to consider these challenges and to map them closely against the agenda established by the NWI. They can be interpreted as a caution, that the institutional change agenda in the NWI is difficult; as realistic, in that institutional change for sustainability is hard; or as a positive endorsement of the NWI as an internationally notable statement of intent by governments and society to move from rhetoric to actual management, policy and institutional change.

A report by the House of Representatives Standing Committee on Environment and Heritage on catchment management (House of Representatives 2000) provides a useful analysis of many of the challenges to implementing a water management reform program such as the NWI. Its neglect since then is an extraordinary example of policy amnesia. The report was produced by an all-party group and was based on months of investigation, extensive consultation and a wide range of submissions. The terms of reference were comprehensive, covering:

- the development of catchment management in Australia
- the value of a catchment approach to management of the environment
- best practice methods of preventing, halting and reversing environmental degradation in catchments, and achieving environmental sustainability
- the role of different levels of government, the private sector and the community in the management of catchment areas

- planning resourcing, implementation, coordination and cooperation in catchment management, and
- mechanisms for monitoring, evaluating and reporting on catchment management programs, including using those reports for state of the environment reporting, and opportunities for review and improvement.

The report discussed the respective roles of the Commonwealth and states, and concluded that the Commonwealth should take the lead role in developing a new system because it was the only government in a position to coordinate and arbitrate competing state interests. The report offered a large number of recommendations; examining them in detail would be a good starting-point for investigating the institutional implications of the NWI.

The report also included consideration of institutional issues at the regional level. This issue is not discussed explicitly in the NWI but it is difficult to conceive of any implementation scenario in which they do not play a leading role. Fifty-six regional natural resource management bodies are now formally in place across Australia (ACIL Tasman 2005), each with a community-based board of management, responsible for integrated management of the region's natural resources. While many started in response to demand from communities (bottom-up), they have been increasingly moulded, homogenised and professionalised to deliver (top-down) programs on behalf of state/territory governments and especially the Commonwealth (Dore & Woodhill 1999; Peters 2006). Regional bodies are already charged, and many commentators would suggest overcommitted, with the regional delivery of two major national resource management programs: Natural Heritage Trust Extension and National Action Plan for Salinity and Water Quality. Most organisations also manage a broader program of activities, encompassing local, regional and other state/national initiatives.

These arrangements differ significantly between jurisdictions, as do the management capabilities of individual organisations. The addition of responsibilities for planning and implementation under the NWI could buckle an already seriously stressed system. That said, the current regional arrangements are widely perceived as the preferred approach to natural resource management delivery (Sinclair Knight Merz 2006), with only few detractors (ITS Global 2006a; Lane et al. 2004).

3 Adoptable knowledge

3.1 Potential Commonwealth–state partnerships

In recent years Commonwealth and state constitutional powers have increasingly overlapped, leading to calls for greater clarity and division of responsibilities. Political scientist Brian Galligan disagrees. According to him, this sharing of responsibility is a positive and fundamental feature of the federal system that promotes consultation and democratic behaviour (Galligan 1995). Galligan argues that the many instances of apparent duplication are not accidents or mistakes that should be tidied up, as the advocates of coordinate federalism argue, and that effort is better invested in making these overlapping institutions more effective.

An indication of the potential for negotiated political solutions that would give both the Commonwealth and the states substantial roles in the development of Australia's 56 regional resource management bodies to implement the NWI is provided by recent reports into regional catchment management by the Australian National Audit Office (ANAO 2004) and the Auditor-General Victoria (2003). The ANAO (2004) report on the performance of the national action plan on salinity, a program that depends on regional bodies for its implementation, concluded that:

> at the regional level, strong and concerted action by all stakeholders is required if the program risks are to be effectively managed. In particular, there are substantial residual risks in small, newly established community-based organisations having primary responsibility for delivering challenging outcomes and managing substantial allocations of Australian government funds (ANAO 2004: 18).

In preparing its report, the Office suggested that state agencies and processes could play a useful role in responding to these issues. For more information regarding this issue it directed readers to the Auditor-General Victoria (2003) report. The Victorian report gives some indication of the role that state governments could play in developing a regional catchment management framework significantly funded by the Commonwealth. Examining these reports side-by-side, a number of differences emerge. The analysis and recommendations of the Australian National Audit Office are more generic than those of the Victorian Auditor-General's Office, applying to a mosaic of six state and two territory systems each working to its own legislation. The main mechanism for achieving compliance with national strategies is the provision or withholding of funds. Managing such programs in detail is difficult for Commonwealth agencies, as they do not have the detailed knowledge, experience and supervisory capacity required although this situation has been improved somewhat by the recent creation of a national facilitator network.

The report of the Victorian Auditor-General, on the other hand, reveals the potential for more detailed supervision created by the involvement of a state government agency. It shows that state governments can be used to exert an additional level of influence on the burgeoning Commonwealth/regional catchment relationship. The recommendations of the Victorian report are given in a context where the state government has legislative control over many aspects of the regional catchment management system. This allows it to apply a wide range of regulatory measures and pressures to clarify responsibilities, improve governance and monitor the performance of boards.

3.2 Experience in other industries

The NWI is an attempt to create a regulatory regime that will promote multi-use of natural resources by different agents operating in the same geographical space. As such it is not unique. The use of leases in relation to land, which dates back to the early 19th century in Australia, is just one example of the way in which governments can implement management policies with multiple objectives. The early pastoral leases were designed to protect long-established Indigenous hunting rights while allowing pastoralism at the same time in the same areas (Holmes 2000).

A more recent example of policy designed to balance competing objectives – environmental protection and timber-based industry – is the Regional Forest Agreement process (e.g. Mobbs 2003). Its stormy record of implementation during the 1990s could provide an indication of what is ahead for the NWI. Regional forest agreements are meant to protect forest areas of high conservation value and provide the timber industry with security of supply – aims very similar to those of the NWI. Eventually the agreements lost credibility with environmental groups, who have increasingly focused their efforts on other options for political action. There has been a failure to effectively involve regional communities, and balancing environmental and production interests has proved difficult. As a result state governments, although often sympathetic to the timber industry, have been unable to secure supply in the face of growing public anger. They also failed to satisfy their environmental critics, resulting in an increasingly polarised debate. This highlights the real difficulty in achieving such compromises, and the dangers that can result from failure to effectively manage politically contentious environmental issues (Mobbs 2003; Hutton & Connors 1999).

Fish are another resource sharing many of the characteristics of water. The size of the available resource is hard to calculate, and often little is known about the reproductive patterns or habitat requirements. Defining a sustainable level of fishing is often only done after industry collapse shows that exploitation was too extreme. Given the level of biophysical knowledge, the precursors to such events are not often easy to pick. This makes fish stocks extremely prone to over-exploitation, because it is hard to set limits that can be sustained in the face of strong political pressure.

Managers of water resources have to grapple with similar issues. The NWI states that, in a given hydrological system, the consumptive pool available for diversion is the water left over after sufficient provision has been made to maintain environmental sustainability at whatever level of modification was agreed in the negotiations required to prepare the water plan. The water resource is obviously finite but it is very hard to define its size, not least because it varies year to year. It is also difficult to fine-tune management given the long lead times between actions and consequences. Failure to successfully juggle the variables involved, however, leads to environmental decline, erosion of resource security and, in many cases, threshold changes that are difficult to predict or reverse.

Fish stock management provides useful lessons for water managers beginning to grapple with the use of property instruments as incentive mechanisms to encourage the holders of water entitlements to manage the resource sustainably. This normally involves proposals to create a package of rights and responsibilities that will shape management behaviour. The aim is to unleash the energy that can be created by self-interest and at the same time channel it to benefit society as well as the individual. It is easy to state but difficult to execute. Managers of fish stocks have been experimenting with this approach for a number of decades. A wide range of variations have been trialled and their successes and failures are well-documented (Grafton et al. 2007).

Placing water management alongside forest and fish management draws attention to underlying continuities and commonalities that are particularly relevant to the process of defining sustainability. The NWI requires that by the end of 2007 all overallocated catchments must have water plans that identify how they will be restored to sustainability. By the end of 2009 all other catchments must have water plans that show how they will avoid overallocation and maintain sustainability. Discussing these issues in the context of similar debates about forests and fisheries highlights a number of general principles. It seems that two conditions provide a minimum that needs to be satisfied before a modified environmental system can be defined as sustainable and resource security protected: the biophysical condition must be stable from a system-wide perspective, and its condition should be politically acceptable to the wider community (Connell 2007).

A system-wide perspective provides a rigorous context for trade-offs between social, economic and environmental demands. Without it, social and economic considerations will usually be privileged over environmental considerations. The second requirement, that of political acceptability in the wider community, creates an opening for the social interests that lack economic power to influence the negotiated definition of sustainability in a given situation. The capacity of such groups to disrupt arrangements from which they are excluded means that they need to be taken into account if resource users want a stable environment for investment. There are many examples of powerful production interests attempting to ignore this dimension of sustainability, not least in the fishing, timber and water industries.

4 Research gaps

4.1 Revisiting recommendations on catchment management

A useful starting-point for a discussion of research gaps relevant to the institutional dimension of the NWI is the House of Representatives Standing Committee Environment and Heritage (2000) report into catchment management. Leaving aside the specifics of what is proposed, its 26 recommendations provide a good checklist of themes that can usefully be investigated, including:

- feasibility of introducing a long-term levy like the Medicare levy to fund environmental rehabilitation
- creation of a seamless organisational structure extending from the national centre to on-ground implementation
- processes to achieve compatible legislation, policy and management across jurisdictions
- information-sharing between jurisdictions and publicly funded organisations
- a national education program to promote understanding of ecological sustainability and the principles of catchment management with a particular emphasis on long-term perspectives
- development of a more effective water entitlements regime
- investigation of all disincentives to ecologically sustainable practices, and
- tax reform.

An overarching recommendation was regional-scale coordination of all funding programs, such as the National Action Plan for Salinity and Water Quality, that affect catchment management.

4.2 Understanding the implications of a rights-based system

There is a need for a wide-ranging and thorough investigation of the implications of replacing the administrative system of water management with a rights-based regime. Much higher standards of data and definitions are now required because decisions may have to survive hostile testing in court. Although only a small percentage of transactions will involve legal action, those cases and the threat of action will shape the definition of rights and responsibilities. As Fisher explains (Chapter 9, this volume), the new system being established under the NWI is based on the assumption that environmental sustainability and resource security will be defined by statute in a form that will be defendable in courts.

However, as Gardner and Bowmer explain (Chapter 4, this volume), none of the nine jurisdictions involved with the NWI have yet incorporated the requirement to achieve environmental sustainability into legislation let alone the supporting detail that will be required to make good on the general principle. The closest that any jurisdiction has come to this minimum standard, they argue, was the NSW *Water Management Act 2000* but even that achievement has been significantly eroded by amendments.

The need to review all legislation, policy and management practices in the nine jurisdictions to remedy these defects and provide a secure legislative foundation for implementation of the new approach to water management was recognised in paragraphs 13, 14 and 27 of the NWI. This process will need to do more than just reflect thinking as of, say, June 2004 when the NWI was released. Many sections of the NWI show awareness that the definition of the 'best available science', which is to be used to define the point of potential overallocation, will evolve – one of the requirements for the water plans is that they incorporate a capacity for progressive renewal. Thus the process of reviewing legislation and policy etc. must be ongoing and, combined with the need to maintain consistency across jurisdictions, will require some sort of interjurisdictional institutional form.

4.3 Increasing the biophysical knowledge base

A closely related issue is the importance of comprehensive biophysical knowledge about the full range of matters relevant to effective water management. Management of hydrological systems is becoming ever more complex. The process of listing relevant factors and working out how best to monitor and document them cannot be dealt with in this chapter. Our task is to pose the question: what sort of cross-jurisdictional organisational structure would be required to undertake this task over an extended time period? National 'state of the environment' reports, the National Land and Water Resources Audit and the benchmarking work recently commissioned by the National Water Commission provide a foundation. However, to satisfy the requirements of paragraphs 48 and 49 of the NWI regarding the need to adjust to drought and climate change and develop the definition of sustainability between now and 2014 using the best available science, a more long-term process and tighter integration into the policy management cycle will be required.

4.4 Building more effective regional management organisations

To deliver the national programs already in place, the current regional arrangements will need to work better than they do. Adding further responsibilities will necessitate greater efforts to build organisational capacity. But in order to build capacity we must first know much more about the basic capacity-related characteristics of these geographically, socially and politically diverse regions. The playing field of regional natural resource management bodies and their boards is uneven in terms of resources, time, information and skills. Consistent delivery of national programs, like the NWI, requires more even distribution of regional capacities across the country. These disparities will not right themselves without purposeful intervention.

Resources

A survey of regional management organisations across Australia reveals considerable variation and indicates the emergence of 'have' and 'have-not' regions, and a growing resourcing issue (Robins & Dovers 2007a, 2007b). Current expenditure on capacity-building activities (self-defined by regional bodies) represents a significant proportion of overall national program budgets (27% or $106 million of National Action Plan for Salinity and Water Quality/Natural Heritage Trust allocation to June 2005). Yet the total proportion allocated to capacity-building between regions ranges very widely, from 0% to 96% of total budgets ($0–7.03 million). Building more effective regional organisations necessitates closer examination of this expenditure and greater attention to achieving minimum capacity levels across all regional bodies and boards.

The base funding of regional organisations varies considerably. Regional bodies in Victoria receive around $1 million per year compared to $250 000 per year in WA. In addition, substantial differences exist in the remuneration of board members, both within and between jurisdictions, with chairs receiving $50 000 per year in NSW but $16 000 in Victoria (Walter Turnbull 2005). Gross funding disparities for individuals with similar responsibilities has the potential to create divisiveness and adversely affect the retention rates of community representatives and program outcomes.

Time

One of the greatest limitations of regional bodies and boards is their time available to meet an increasing list of responsibilities. Capacity-building measures cannot make time, but they can help ensure that the available time is used well. Regional organisations have long called for more streamlined administration and reporting (WalterTurnbull 2005; ITS Global 2006b). The Natural Resource Management Ministerial Council established the Red Tape Reduction Taskforce in April 2004 in response to these concerns, but it did not achieve 'the desired impact on streamlining administrative processes for regional bodies' (ITS Global 2006b: 28). Administrative and reporting requirements need to reflect organisational differences, and not create a disproportionate burden for smaller entities. The current level of duplication of effort across regions is unnecessary and excessive, reducing the finite time that regional actors have to commit to resource management planning and implementation.

Information and advice

Governments have yet to make substantive progress as providers of scientific information to regional bodies and boards, who remain overwhelmed with information of variable relevance and accessibility (Robins 2006), and have limited access to scientific and technical input (Davis et al. 2001). Campbell (2006b) concluded that 'it is too hard for people in any part of the system to find out what is happening and what is being learned elsewhere – or has been learned already'. Sinclair Knight Merz (2006: 3) found widely variable science underpinning salinity responses between regions and that 'use of best-practice "models" for salinity intervention was not well documented in the NRM plans'. There is much scope for cross-regional harnessing of local expert knowledge (Whelan & Oliver 2005; Campbell 2006a), not only to inform regional actors but also government agencies. Better structures for information and knowledge management and exchange between jurisdictions and regional bodies are needed (ITS Global 2006b).

Skills

Skills development is a high priority for regional management organisations. Capacity-building measures for addressing human, social (cognitive and structural), institutional and economic capital are outlined in an options discussion paper, which draws ideas from the resource management, health and risk and emergency management literatures (Robins 2007a). Some examples of skills-related strategies that are worth exploring further include establishing a training facility and formalising debriefing processes. Currently, great variation exists between jurisdictions with induction and ongoing training of regional staff and board members. Training for board members in statutory jurisdictions tends to be centrally funded, whereas non-statutory based organisations are generally self-funded. Minimum requirements for board training, resources for training and board remunerations are issues that could be standardised through bilateral agreements. Savings could be achieved through reduced duplication of effort and greater cross-jurisdictional information-sharing.

Management of hydrological systems is much more complex than it was only a few decades ago. Some of the skills required to deal with these issues were listed in 2002 by David Dole, then General Manager of River Murray Water. He explained that the future water manager will need:

- technical knowledge of the hydrology and hydraulics of whole river systems including their floodplains
- technical knowledge of whole catchment land/water processes
- technical engineering skills relevant to constructing, operating and managing physical works
- understanding of the biophysical relationships between water, land and environment, including the impacts of changing flow regimes on river ecosystems

- understanding of the water needs of natural systems as well as those of consumptive users
- technical skills in improving the efficiency and effectiveness of the processes that convey water from storage to root zone
- technical skills in managing and treating drainage waters and in achieving effective surface or subsurface drainage
- a commitment to creating sustainable natural resource systems while also achieving reasonable economic outcomes
- the ability to work with communities to jointly build a sound knowledge base to underpin the negotiation of future actions
- the confidence to recognise the limits of current knowledge of social impacts on natural systems and the integrity to recognise and promote the need for change (Dole 2002: 35).

Under the NWI, we could add the need for a good working knowledge of water law and policy.

Where will these new super water managers come from? Future demand will be much greater than now, but there are already serious shortages of appropriately skilled people. This personnel gap at the regional level is emerging at the same time as similar shortages are becoming serious in many parts of the water management system. It parallels developments in other spheres of Australian life. Whether the subject is social welfare, transportation, medicine, engineering, business or sports administration, the level of skills required has increased dramatically in recent years. Skill and human resource shortages could well prove the greatest source of risk to the NWI and Australian water management in the medium term. This increases the pressure not only to find more people and train them to a higher level but also to design the most efficient possible system of water management. Compromises that increase transaction costs not only squander scarce resources, they also jeopardise the likelihood of effective implementation of the NWI.

Conclusion

This chapter has argued that the institutional reform agenda implicit and explicit in the NWI is very significant, probably more than many in the policy community understand it to be. Particular challenges lie in national–state–regional institutional coordination, information quality and delivery, statutory reform, capacity-building in regional organisations, and development of the skills and human resources needed to run a more complex and exacting water policy and management system.

As startling as the scale of these challenges are, they are nonetheless to be expected, given the nature of the NWI as an innovative and far-reaching proposal for implementing, in the Australian water context, the difficult and long-term idea of ecologically sustainable development. It is an institutional agenda worthy of pursuing with careful thought, vigour and intent.

References

ACIL Tasman (2005). *Institutional arrangements in the Australian water sector*. National Water Commission: Canberra.

Auditor-General Victoria (2003). Part 3.9: Sustainability and Environment. In *Report on public sector agencies – results of special reviews and 30 June 2003 financial statement audits*. Melbourne.

Australian Bureau of Statistics (2005). *Informing a nation: the evolution of the Australian Bureau of Statistics*. ABS: Canberra.
Australian National Audit Office (2004). *The administration of the national action plan for salinity and water quality*, Australian National Audit Office: Canberra.
Campbell A (2006a). *The Australian natural resource management knowledge system*. Land and Water Australia: Canberra.
Campbell A (2006b). The NRM knowledge system in Australia. *RipRap: River and Riparian Lands Management Newsletter* **30**, 7–8.
Connell D (2007). *Water politics in the Murray-Darling Basin*. Federation Press: Sydney.
Connor R & Dovers S (2004). *Institutional change for sustainable development*. Edward Elgar: Cheltenham.
Council of Australian Governments (1994). *Attachment A: A water resource policy*. February, Canberra.
Cullen P (2006). Water planning panel. Speaking notes. 4 September, Brisbane Riversymposium.
Davis JA, Finlayson BL & Hart BT (2001). Barriers to science informing community-based land and water management. *Australasian Journal of Environmental Management* **8**, 99–104.
Dawes W, Gilfedder M, Walker GR & Evans WR (2004). Biophysical modelling of catchment scale surface and groundwater response to land-use change *Mathematics and Computers in Simulation* **64** (1), 3–12.
Dole D (2002). *Managers for all seasons: uncharted waters*. Murray–Darling Basin Commission, Canberra.
Dore J & Woodhill J (1999). *Sustainable regional development. Final report: an Australia-wide study of regionalism highlighting efforts to improve the community, economy and environment*. Greening Australian: Canberra.
Dryzek J (1996). The informal logic of institutional design. In *The theory of institutional design* (ed. RE Goodin), pp. 103–125. Cambridge University Press: Cambridge.
Galligan B (1995). *A federal republic: Australia's constitutional system of government*. Cambridge University Press: Melbourne.
Grafton RQ, Kompas T, McLoughlin R & Rayns N (2007). Benchmarking for fisheries governance. *Marine Policy* **31**, 470–479.
Holmes J (2000). Pastoral lease tenures as policy instruments, 1847–1997. In *Environmental history and policy: still settling Australia* (ed. S Dovers), pp. 212–242. Oxford University Press: Melbourne.
House of Representatives Standing Committee on Environment and Heritage (2000). *Coordinating catchment management: report of the inquiry into catchment management*. December, Federal Parliament: Canberra.
Hutton D & Connors L (1999). *A history of the Australian environment movement*. Cambridge University Press: Melbourne.
ITS Global (2006a). *Evaluation of the national investment stream of the Natural Heritage Trust of Australia*. Final report. ITS Global: Melbourne.
ITS Global (2006b). *Evaluation of the bilateral agreements for the regional component of the Natural Heritage Trust of Australia*. Final report. ITS Global: Melbourne.
Lafferty WM (ed.) (2004). *Governance for sustainable development: the challenge of adapting form to function*. Edward Elgar: Cheltenham.
Lane MB, McDonald GT & Morrison TH (2004). Decentralisation and environmental management in Australia: a comment on the prescriptions of The Wentworth Group. *Australian Geographical Studies* **42:** 103–115.
Lenschow A (ed.) (2002). *Environmental policy integration*. Earthscan: London.

Mobbs C (2003). National forest policy and regional forest agreements. In *Managing Australia's environment* (eds S Dovers & S Wild River), pp. 15–34. Federation Press: Sydney.

Morley M & Thompson A (2006). Knowledge for regional NRM: connecting researchers and regions. *RipRap: River and Riparian Lands Management Newsletter* **30**, 11–13.

Murray-Darling Basin Ministerial Council (1995). *An audit of water use in the Murray-Darling Basin: water use and healthy rivers – working towards a balance.* June. Murray-Darling Basin Ministerial Council, Canberra.

National Water Commission (2006). *Distilled* (monthly newsletter), October.

Peters M (2006). Towards a wider debate on federal and regional governance. In *Federalism and regionalism in Australia: new approaches, new institutions? A national symposium*. NSW Farmers Association and Griffith University. New Parliament House: Canberra..

Robins L (2006). A model for knowledge transfer and adoption: a systemic approach to science communication. *Environmental Science & Policy* **9**, 1–9.

Robins L (2007a). *Enabling Regional NRM Boards: a discussion paper on capacity building options.* Centre for Resource and Environmental Studies, Australian National University: Canberra.

Robins L (2007b). Major paradigm shifts in NRM in Australia. *International Journal of Global Environmental Issues.* In press.

Robins L & Dovers S (2007a). Community-based NRM boards of management: are they up to the task? *Australasian Journal of Environmental Management.* In press.

Robins L & Dovers S (2007b). NRM regions in Australia: the 'haves' and the 'have nots'. *Geographical Research* **45** (3).

Sinclair Knight Merz (2006). *Evaluation of salinity outcomes of regional investment: final report.* Prepared for Departments of Environment and Heritage/Agriculture, Fisheries and Forestry. Sinclair Knight Merz: Bendigo, Victoria.

Van Dijk AIJM, Austin J, Theiveyanathan R, Benyon PB & Hairsine PB (2004). *Representing plantation scenarios in hydrological modelling.* CSIRO Land and Water technical report. CSIRO Land and Water.

Van Dijk A, Evans R, Hairsine P, Khan S, Nathan R, Payday Z, Viney N & Zhang (2006). *Risks to the shared water resources of the Murray-Darling Basin.* Murray-Darling Basin Report: Canberra.

Walter Turnbull (2005). *Evaluation of current governance arrangements to support regional investment under the NHT and NAP.* Departments of Environment and Heritage/Agriculture, Fisheries and Forestry: Canberra.

WCED (World Commission on Environment and Development) (1987). *Our common future.* Oxford University Press: Oxford.

Whelan J and Oliver P (2005). Regional community-based planning: the challenge of participatory environmental governance. *Australasian Journal of Environmental Management* **12**: 126–135.

Young OR (2002). *The institutional dimensions of environmental change: fit, interplay, and scale* MIT Press: Mass.

Chapter 11

International perspectives on water policy and management

Emerging principles, common challenges

Karen Hussey and Stephen Dovers

Humanity's disruption of the hydrological cycle has been immense, particularly in the last century when our demand for water increased six-fold (UN 1999). In that time, the profligate use and abuse of this natural resource has produced serious problems in terms of both the availability of adequate amounts of freshwater, and ecosystem health.[1] Water is a basic human need and key component of human development, so disruption of the hydrological cycle is a necessary by-product for our existence. In recent years, though, there has been increasing emphasis placed on the important function played by freshwater resources as an ecosystem service. Through their role in the hydrological cycle, rivers, lakes and underground aquifers provide a renewable source of freshwater for irrigation, households, industries and other uses that require the removal of water from its natural channels (Daily 1997: 197). From this perspective, the relationship between freshwater resources and human development is manifest:

> The challenge in this new situation is to actively enhance and strengthen the capacity of the biosphere to support and sustain social and economic development and to explicitly recognise the role of freshwater in this context (Rockström et al. 1999).

Broadly, policy and management interventions must seek to influence highly complex, interdependent social and ecological systems in a manner that appreciates this (Folke 2006). The development of Australia's National Water Initiative (NWI) recognises the need for such interventions and the legal, institutional and political requirements to achieve them. Nevertheless, in a complex policy domain such as water, the tensions in trying to balance the range of values held by the multiple players in the policy community have meant that, for every success, compromises have been made and residual implementation challenges remain (Connell et al. 2006).

The challenges inherent in water policy are not isolated to Australia and many developments have taken place internationally which reflect our own experiences. While the environmental and socio-economic conditions vary across regions, there are common problems in relation to institutional and regulatory fragmentation in diverse political systems, reconciling environmental needs for water with consumptive demands, the capacity of newly emergent catchment (or river basin) institutions to successfully fulfil their water policy responsibilities, and difficulties in choosing appropriate policy instruments to meet varied objectives. To inform Australia's implementation of the NWI, it is useful to place our experiences in the context of international developments. This chapter offers both positive and negative lessons for Australia's water reform agenda, from approaches to water policy and management in the European Union (EU), Canada, South Africa and the United States (US).

In the first section we provide a brief background on the key principles of the NWI and the water policies of the four case studies. In Section 2, we explore the different approaches adopted in relation to 'environmental water', that is, the attempt to reconcile consumptive and environmental water demands. In Section 3, we examine the challenges faced by federal systems in establishing appropriate institutional and regulatory arrangements to account for diversity across regions and different levels of government, in particular the place of catchment-based institutions as crucial new institutions. In Section 4, we analyse the choice of policy instruments in each country, which proves particularly interesting in light of the NWI's emphasis on water markets. In Section 5, we draw on these developments to highlight the positive and negative lessons relevant to the implementation of the NWI.

1 Emerging principles in water policy and management

The development of the NWI represents a multi-component, national-level policy framework of unprecedented scale – a 'new philosophical approach to water management' (Connell et al. 2005: 83). It is a major step forward in Australian water policy: first, because important proposals and ideas now have official status at national level and, second, because the NWI is now the primary statement of policy with which all water-related legislation and policy must comply.[2] Within the NWI, it is possible to identify a number of core imperatives; none are peculiar to Australia but the combination perhaps is, especially in the context of extreme climate variability and the specific constitutional and legal system (see Hussey & Dovers 2006).

- Security and predictability of water entitlements for extractive users, in particular irrigated agriculture. It is useful to separate two related yet different imperatives here, which at times are conflated and lead to confusion. The first suggests public subsidy and long-term planning, the latter the primacy of efficiency and market forces:
  – the social and equity imperative of maintaining water-dependent communities, established purposefully through previous government policy
  – the economic imperative of productivity, competition in export trade and efficiency in inputs (including of policy interventions).
- Increasing prospects of water scarcity and the need for efficiency of use and change to allocation rules, a product of increased extraction, greater number of water users (including environment), climate change and variability, and physical and economic limitations to expanding supply.
- Widening of water policy and management debates to create a national water policy arena (as opposed to several, state- or catchment-limited arenas) and greater awareness of urban and peri-urban as well as rural water issues.
- An increasing role for catchment and regional organisation and delivery of policy objectives, viewed as an efficient and suitable scale for water management.
- The importance of participatory approaches to resource management, at the catchment/regional scale but also at local (or district) scale through numerous programs epitomised by Landcare. (The allowance of public participation in higher-order policy formulation is not so apparent.)
- Ongoing efforts to rationalise or at least coordinate roles and responsibilities for resource management within a federal system. The main debates concern the Commonwealth versus state/territory powers, with the role of local government often ignored (Wild River 2003). Whatever the relative power acceded to the Commonwealth

as a result of current political negotiations, water will remain a shared policy responsibility across governments.
- Increased emphasis on environmental water flows, not simply in volume but also quality and timing of flows to maintain riverine ecosystems.[3]
- Recognition of Indigenous water values and uses, encompassing environmental, cultural, subsistence, health and economic dimensions.
- As in all areas of public policy, changes in policy approaches and management styles influenced by dominant neo-liberal political philosophy and neo-classical economic theory, manifesting in the Australian term 'economic rationalism', and managerial styles termed 'new public management' (McLaughlin et al. 2002).

As illustrated by the case studies in this chapter, these principles are not unique to Australia although the approaches – and combination of approaches – taken differ between regions. It is through these differences that comparative policy insights can be identified.

The case studies were chosen because of shared characteristics with Australia: all the case studies labour under the challenges of multi-jurisdictional and multi-level governance arrangements, especially in the EU; all are characterised by a strong and productive agricultural sector which share resources between states for agricultural and industrial production; South Africa, the west coast of the US and the southern states of the EU share Australia's water quantity issues; all experience problems with water quality; and, finally, all are characterised by relatively stable liberal-democratic governments which have the financial and political means to tackle water management issues. With the exception of the US, all have undergone significant water reforms in recent years. Space prevents us from detailing the often-tortuous process through which contemporary water policy has emerged in each, but it is possible to map the institutional and policy styles adopted, as well as the provisions made for some of the issues discussed in the other chapters in this volume, such as Indigenous issues and public participation (Table 11.1).

1.1 Reconciling environmental needs for water with human needs

Reflecting growing public concern for the environmental health of Australia's water resources, the NWI requires the complete return of currently overallocated or overused surface and groundwater systems to environmentally sustainable levels of extraction (para. 23(iv)). Water that has been identified by the states and territories to meet agreed environmental and other public benefit outcomes is to be given statutory recognition and 'at least the same degree of security as water access entitlements for consumptive use' (para. 35(i)), and only water that is not required to meet environmental outcomes can be made available for trading (para. 35(iii)). In this way, the NWI is a major step forward. There are, however, a number of significant challenges in implementing these objectives. The first relates to the definition of environmentally sustainable levels of extraction. This term is defined to mean the level of water extraction from a particular source which, if exceeded, would compromise key environmental assets or ecosystem functions and the productive base of the resource (NWI B1: 29). In recent years considerable research effort has focused on the connections between surface and groundwater systems (Dept of the Environment and Heritage 2004),[4] the impacts of various land use practices and interception (Dillion et al. 2001; Keenan et al. 2004), and the merits of various approaches to environmental management (Dovers 2005; Gunningham 1999; Gunningham & Sinclair 2004). While progress has been made, the complexities that exist, particularly in dynamic conditions such as under climate change, in response to bushfires or as a result of different land uses and tenures, suggest that defining environmentally sustainable levels of extraction will require far more research than is currently advocated in the NWI.

A second problem relates to the number of regional water systems that will need significant management adjustments to achieve that standard. It is the responsibility of the states and

Table 11.1 International approaches to water resource management

	Australia	EU	US	South Africa	Canada
Water problems	Predominantly water quantity, and salinity problems as derivative	Predominantly water quality, but quantity issues in southern European states such as Spain	Varies: west coast quantity, east coast quality	Quality and quantity – major problems with both. Environmental conditions similar to Australia (arid and semi-arid)	Quality, but emerging quantity issues in urban and peri-urban sectors in southern provinces
Water policy framework	National Water Initiative	Water Framework Directive	State-based; and the *Clean Water Act*	National Water Bill	Federal water policy
Central coordinating unit	National Water Commission	No – managed by DG Environment. Member states to implement	No – severely fragmented	Federal Department of Water Affairs and Forestry	Federal Interdepartmental Committee on Water
Provision for environmental water	Environmental water allocations requirement in the NWI	'Good ecological status' required, but policy instrument choice up to the member states	Yes, state-based and varies considerably between them	Ecological – component of the reserve kept back to meet environmental objectives	Environmental flow regimes in place at provincial level
Institutional arrangements	Catchment-based institutions	River basin institutions and plans	Recent push for river basin management but not very advanced. Mostly state-based	19 newly established catchment-based institutions	Managed by provinces, in cooperation with federal and local governments
Choice/style of policy instrument	Water trading, pricing	Water pricing (full cost recovery)	Water trading (west coast), pricing	Water pricing, exploring tradable water rights	Polluter-pays principle and water pricing
Recognition of Indigineous interests	Inclusion of Indigenous interests in water planning, where possible. Native title considered in water plans	No – only provision is for public participation and consultation	American Indian and Alaska Native Tribal Government policies – recognised native title rights to water. 1996 amendment to CWA recognised qualifying tribes as states	Access to safe drinking water a human right. Key objective of NWB to 'redress the results of past racial and gender discrimination'	First Nations Water Management Strategy – main focus on safe drinking water in Indian and Inuit communities

territories to develop water plans that will secure ecological outcomes for water systems (para. 37(i)) and resource security outcomes for consumptive use (para. 37(ii)). However, the areas to be covered, level of detail required, duration and frequency of review, and the amount of resources devoted in each plan is determined exclusively by the state or territory (para. 38), and reading of the NWI suggests that formal identification of currently overallocated water bodies is likely to be limited (Connell et al. 2006: 87; NWI paras 33(ii) and 43).

The primary means through which environmental and other public good outcomes are to be achieved in the NWI is through environmental water allocations (EWAs), and establishment of accountable environmental water managers[5] with the necessary authority and resources to provide sufficient water for EWAs. However, much remains to be done to achieve nationally the NWI objectives for environmental and other public goods. Three particular factors are proving problematic (see Chapters 4, 9 and 10, this volume):

- adequacy of scientific information and transparency of the process for determining ecological objectives and environmental allocations
- considerable variation between jurisdictions in the expression of the legal and policy provisions and in the extent of EWAs provided
- institutional capacity of environmental water managers, whether it be a catchment management authority or new arrangement, in terms of funding, independence, accountability and knowledge base.

The shift toward provision for environmental water is not isolated to Australia. Each country in the case studies provides for the needs of the environment and faces similar implementation challenges. In the EU's Water Framework Directive (WFD), a general requirement for ecological protection and a general minimum chemical standard were introduced to cover all surface waters. These two elements are referred to as 'good ecological status' and 'good chemical status'. Good ecological status is defined in Annex V of the WFD in terms of the quality of the biological community, and hydrological and chemical characteristics. As no absolute standards for EU-wide biological quality can be set, because of ecological variability, the controls are specified as allowing only a slight departure from the biological community which would be expected in conditions of minimal anthropogenic impact. A set of procedures for identifying that point for a given body of water, and establishing particular chemical or hydromorphological standards to achieve it, is provided, together with a system for ensuring that each member state interprets the procedure consistently (CEC 2000: Annex V). One criticism is that the EU system is very complicated, but this is inevitable given the extent of ecological variability, and the large number of parameters which must be dealt with. It is also typical of EU directives, which are traditionally prescriptive and onerous in terms of implementation (Lampinen & Uusikyla 1998; CEC 2003). Indeed, the prescriptive nature of the WFD in terms of the definitions for good ecological status, together with requirements for monitoring and evaluation, may prove counter-productive as the member states move to implementation. However, where the EU's WFD may be too prescriptive, it does have the advantage of direct effect in the member states, which means that incorrect application or lack of compliance with the directive could result in severe financial penalties. In this sense, it carries more weight than Australia's NWI, which relies on positive incentives for implementation through the Water Smart and Raising National Water Standards funds.

In South Africa there is no specific part of the *National Water Act* with explicit provisions for water conservation and demand management, although the Act's definition of conservation makes it clear that water conservation and demand management measures are an essential component of water resource management. Indeed, the Act's fundamental objectives for managing South Africa's water resources are 'to achieve equitable access to water resources and their sustainable and efficient use' (DWAF 2004: 56). To give effect to the interrelated objectives of sustainability and equity, the approach to managing water resources comprises two complementary strategies:

- resource-directed measures, focusing on the quality of the water resource itself, which is ultimately a measure of its ecological status
- source-directed controls, which contribute to defining the limits and constraints that must be imposed on the use of water resources to achieve the desired level of protection, and include such tools as standards, water use licences and authorisations.

Of particular interest in South Africa's policy, in light of Australia's difficulties in determining what constitutes environmentally sustainable levels of extraction, is the flexible water resources classification system being developed 'to provide a consistent framework within which water resources can be classified, each class representing a different level of protection' (DWAF 2004: 57). The classification system comprises four levels: Natural, Moderately Used/Impacted, Heavily Used/Impacted, and Unacceptably Degraded Resources. Unlike the EU and Australian policies, which prescribe a minimum standard (allowing for problems in definition), the South African four-tier classification system is explicit in its integration of social and economic issues in attributing levels of protection: 'it is not possible for all resources throughout the country to be given a high level of protection without prejudicing social and economic development ... [therefore] social and economic considerations will be included in the classification process to ensure a decision that provides a balance between protection and utilisation' (DWAF 2004: 59). Thus, in determining the ecological component of reserve, an integrated approach allows for:

- the quantity and quality of water to meet basic human needs and protect aquatic ecosystems
- resource quality objectives such as quantity, pattern and timing of instream flow, water quality, the character and condition of riparian habitat etc.
- user requirements and the class of the resource.

This does not differ too much from Australia's approach in so far as the requirements to meet 'healthy river status' in the Murray-Darling Basin was deemed by the Scientific Advisory Committee to involve the return of 1400 GL to the river. However, the social and economic costs of doing so reigned supreme and the final figure agreed upon in the Living Murray Initiative was 500 GL (Jones et al. 2002).

The NWI places overwhelming emphasis on the environmental effects of overallocation and overuse of water resources vis a vis water supply. There is little mention of water quality, other than reference to the National Action Plan for Salinity and Water Quality (and it could be argued the water quality referred to in that context – salinity – is a derivative of water quantity issues), the National Water Quality Management Strategy, and brief mention of water quality as one of the 'environmental and other public benefit outcomes' to be achieved more generally through the NWI. The lack of attention to water quality in the NWI is worrisome given the increasing problems experienced by Australian urban centres, particularly Adelaide and Sydney (Smith 2001: ch. 2). In contrast, water quality receives far more attention in South Africa, the EU and Canada.

Such a balance is perhaps understandable in the EU, owing to the high level of intensive agricultural production, high-density living and significantly older water infrastructure than in Australia. In Article 1 of the WFD, water quality is a stated priority such that the purpose of the WFD is to establish a framework for the protection of inland surface waters, transitional water, coastal waters and groundwater which:

- aims at enhanced protection and improvement of the aquatic environment, inter alia, through specific measures for the progressive reduction of discharges, emissions and

losses of priority substances and the cessation or phasing-out of discharges, emission and losses of the priority hazardous substances
- ensures the progressive reduction of pollution of groundwater and prevents its further pollution.

The WFD also explicitly states the relationship between other EU water quality-related directives, including the nitrates, urban wastewater, and integrated pollution prevention and control directives. Furthermore, where a quality objective or standard requires stricter conditions than those which would result from the application of any of these directives, the more stringent controls are enforceable (WFD Article 10(3)).

Until very recently, the emphasis in Canadian water policy has been on water quality, with water quantity issues only emerging in the mid 1990s, particularly with concerns for the impact of climate change. The federal government has identified two main goals for water quality and quantity. The first goal relates to the need to protect and enhance the quality of the water resource, and states that more stringent regulations and standards alone cannot protect water resources without economic incentives/penalties to prevent their impairment (Environment Canada 2003).

As already stated, the US has not developed a national overarching framework for water policy and management. It suffers from significant fragmentation both horizontally across land and water law and vertically across different levels of government (see below; Andreen 2006). In relation to water quality particularly, there is national legislation in the form of the *Clean Water Act*, although it is focused primarily on industrial point source pollution. For non-point source pollution, such as from agriculture, the US Congress instructed states to develop water management plans but the policy was watered down to permit the states to use exclusively non-regulatory avenues such as technical assistance, education, training and demonstration projects to implement management plans (Andreen 2006: 10283).

Therefore, while the NWI does not address water quality issues to the extent that it perhaps should, and indeed to the extent the other countries do, the existence of Australia's National Water Quality Management Strategy (1992) allays any fears that Australia's policy is as fragmented and ineffective as that in the US.

2 Institutional arrangements for water resource management

The success of the NWI relies heavily on coordination of the multiple levels of government within the federal system. At Commonwealth level, responsibility for overseeing the implementation of the NWI lies with the Natural Resources Management Ministerial Council (paras 18 and 104) with significant support from the National Water Commission (para. 19), a federally funded body whose Commissioners are dominated by the Commonwealth government.[6] The development of detailed water plans are the responsibility of the states and territories and the efficacy of a national trading scheme will depend on the coordination between states as much as between the states and Commonwealth. However, as evidenced by the 100-year-old Murray-Darling Basin arrangements, protecting state sovereignty remains a high priority in the Australian federal system (Connell et al. 2005: 89). The only tangible incentive the Commonwealth has to encourage state cooperation is the $A2 billion in funding that states can access in return for compliance. However, even where intergovernmental coordination is politically sanctioned by the states, the transaction costs incurred with so many jurisdictions and players involved remains high. Finally, it is implicit in the NWI that much of its implementation will be driven by regional catchment management authorities. A catchment management approach to water resources management is not new in Australia (see Murray-Darling Basin Commission 2002) but the NWI makes their role far more ambitious, complex and politically contentious (Connell et al. 2005: 89).

All of the countries considered share the challenges of multi-jurisdictional and multi-level governance arrangements, which in three countries have encouraged the adoption of national-level policy frameworks. Furthermore, just as the NWI established the National Water Commission, centralised coordinating bodies have also been established under South Africa's National Water Bill (the Department of Water Affairs and Forestry) and Canada's Federal Water Policy (the Interdepartmental Committee on Water). However, unlike the National Water Commission, Canada's Interdepartmental Committee does not have significant funding capability and therefore lacks the same level of incentive with which to persuade the provinces to comply. In South Africa, the DWAF is mandated under the *National Water Act* with the power to regulate the use, flow and control of all water in the republic across the nineteen catchments (DWAF 1998).

Like Australia, Canada is a federation and different levels of government have different jurisdictional roles related to water management. Canadian provinces and one of the territories have primary jurisdiction over most areas of water management and protection, although most of those governments delegate certain authorities to municipalities, especially potable water supply and urban wastewater management. They may also delegate some functions to local authorities responsible for a particular area or river basin. Most major uses of water in Canada are permitted or licensed under provincial water management authorities. Provincial legislative powers include, but are not restricted to, areas of flow regulation, authorisation of water use development, water supply, pollution control and thermal and hydroelectric power development. Federal jurisdiction applies to the conservation and protection of oceans and their resources, fisheries, navigation and international relations, including responsibilities related to the management of boundary waters shared with the US. Much like the strategies and policies developed under the Council of Australian Governments (COAG), Canada has significant shared federal–provincial responsibilities related to water including agriculture, significant national water issues and health (Environment Canada 2003).

Unlike in Australia, the EU, Canada and South Africa, the US has not developed a national framework for the management of water resources. Water policy and law in the US is significantly fragmented: eastern states are predominantly governed by the doctrine of riparian rights, while the western states have a system that treats water as a kind of private property or commodity, known as prior appropriation (Hussey 2006: 10260; Andreen 2006: 10280). There is a third system known as regulated riparianism which shares characteristics with the Australian legal system, whereby the withdrawal of water requires prior approval and use of permits and licences. There are a number of deficiencies in US water law that have not been addressed nationally and are unlikely to be resolved in the near future. First, there are jurisdictional barriers which have hindered attempts to successfully legislate for both human and environmental needs in the US to date (Arnold 2005). In particular, jurisdictional fragmentation exists both between levels of government and within the same level of government. For instance, while most point source pollution control is federal, through the Environmental Protection Authority, the responsibility for regulating non-point source pollution lies in state hands. Further, wetland regulation is primarily the province of the US Army Corps of Engineers, whereas enforcement of the *Endangered Species Act* in inland waters belongs to the US Fish and Wildlife Service. Moreover, while pollution law and protection of biodiversity reside largely in federal hands, water quantity is a matter of state law – and land use management is generally the domain of local government. As Andreen (2006: 10279) warns, successive US legislators 'have spent decades creating ...

separate legal systems to govern land use, water use, and water pollution, and it will take considerable effort to demonstrate to voters, economic interests, and decision-makers at all levels of government precisely how land use and water are inextricably connected throughout the whole of a watershed'.

Such jurisdictional fragmentation leads to a second significant problem of regulatory fragmentation between water rights and land rights. In the US, there is a considerable problem with coordination both vertically with respect to federal and state water law and local land use management, and horizontally with respect to the various agencies and political entities that have responsibilities within each subject area. As in Australia, there have been attempts in the US to overcome these jurisdictional barriers including the public trust doctrine and the introduction of environmental impact assessments. However, unlike in Australia and the EU – where environmental impact assessment has been incorporated into all state (and member state) legislation – in the US it has only been enacted into fifteen state statutes (Andreen 2006) in addition to the federal statute. In an effort to overcome this multi-layered jurisdictional puzzle there have been calls for the development of watershed institutions, much like the catchment-based institutions in the NWI, WFD and South Africa's National Water Bill. However, to date there has not been a comprehensive federal push for this kind of regime change. As Andreen (2006: 10287) warns: 'Congress would have to be persuaded to act, and federal funds would have to be provided to finance this new organisational structure and its activities. In the current political climate, this is a huge, perhaps insurmountable, challenge'.

Integrated catchment-based approaches to water resource management are thought to be the best means of securing the sustainable supply of crucial ecosystem goods and services on which social and economic development depends (Falkenmark 2003). Interestingly, Australia has paved the way in catchment management with the Murray-Darling Basin Initiative. Since then, almost all jurisdictions (with the exception of the US) have shifted to a catchment-based approach. However, there are two challenges confronting Australia and the case studies:

1. establishing appropriate institutional arrangements to encourage policy coherence and coordination between catchment management authorities, and between those authorities and state and federal departments
2. equipping catchment-based institutions with the necessary financial, human and intellectual capacity to enable the not-insignificant task of implementing water policy (and related policies on biodiversity, land management, planning etc.).

In the EU, there is the added difficulty that some river basins cross member state borders and thus the appropriate institution becomes an international river basin district. The member states must ensure the environmental objectives of the WFD identified in Article 4 are achieved through coordination between institutions and across borders (CEC 2000: Article 3(3–4)). In South Africa, some commentators have expressed concern at the significant difficulties in establishing the new institutional arrangements for catchment management (de Coning & Sherwill 2004 :44):

> It is clear that institutional strengthening of various organisations will be necessary to ensure implementation. A number of areas of research exist such as cooperative government and governance in water, roles and responsibilities of CMAs (important research is already in progress in this area), establishment conditions, the future financial viability of CMAs and other entities, intergovernmental relations and coordination and human resource development.

While Australian analysts share concerns about the institutional capacity of catchment management authorities, particularly under the NWI, Australia is far more experienced in catchment management than other countries are. For that reason, it is probably most likely that the rest of the world will be looking to to learn from us, rather than the other way around.

3 Policy instruments

If the NWI represents, on the one hand, the consolidation of environmental sustainability in water resource management, on the other hand it embraces the Australian governments' preference for policies grounded in neo-liberal political philosophy and neo-classical economic theory to achieve that end – in this case, the use of market-based instruments and property rights for the management of public goods. It is clear that the overarching objective in the NWI, upon which all other objectives lie, is the establishment of clear and nationally compatible water access entitlements to facilitate the operation of water markets within and between jurisdictions (paras 24, 25 and 58(i)). The wholehearted adoption of water markets reflects economic rationality and the imperatives of efficiency, productivity, a limited role for government and, crucially, the need for well-defined property rights: 'environmental problems must be understood more as failures by governments to specify property rights than as offshoots of private profit-seeking' (Mitchell & Simmon 1994: 148).

In analysing the policy instruments adopted in each of the international case studies it is evident that there is a concerted shift towards economic-based instruments in water resource management. In particular, two instruments are favoured: full cost recovery pricing, and water trading. In Canada, the federal government is committed to the concept of 'a fair value for water' (Environment Canada 2003). To implement this concept in federal policies, programs and initiatives, it has:

- endorsed the concept of realistic pricing as a direct means of controlling demand and generating revenues to cover costs
- developed new water-efficient technologies and industrial processes that minimise costs, and encourage water conservation and improved water quality
- encouraged application of pricing and other strategies, such as the beneficiary/polluter-pays concept.

Similarly, an important and central point of the EU's WFD is the establishment of a full cost recovery framework for the water price system (Hansjugens 2002). Article 9 determines 'Member states shall take into account the principle of recovery of the costs of water services, including environmental and resource costs ... in accordance with the polluter-pays principle (CEC 2000). Article 9 thus represents the central regulation of the EU's WFD and requires member states to develop (by 2010) 'water pricing policies [that] provide adequate incentives for users to use water resources efficiently, and thereby contribute to the environmental objectives of this directive'. In addition, price reductions for special customers such as big companies or irrigators are no longer compatible with EU water legislation, and river basin authorities can expect to increase their revenues after the shift to full cost recovery water pricing. However, one interesting question in the EU context is that it is unclear which institution should receive the additional water revenues from water prices, and whether it is possible to transfer the funds collected to other catchment areas for water efficiency or environmental measures (Kuckshinrichs & Schlor 2005). Likewise, South Africa's National Water Bill states the need to adopt a pricing strategy to 'contribute to achieving equity and sustainability in water matters by promoting financial sustainability and economic efficiency in water use' (DWAF 2004: 83).

In addition to pricing policies, there is increasing interest among Canadian policy-makers in the potential of water trading, particularly as Canada suffers from a similar imbalance to Australia between excess water supply in the southern provinces and water scarcity in the northern provinces (although reversed, obviously). Canada's entitlement schemes are fragmented across the provinces, but unlike Australia there are no plans to adopt a nationally consistent approach to entitlements and thus it is unlikely Canada will pursue widespread interstate trading in the immediate term. In South Africa, water trading has been taking place among irrigators along the Lower Orange River and the National Water Bill makes provision for tradable water use entitlements, although it cannot be implemented fully until the compulsory licensing process – currently underway – has been completed (DWAF 2004: 89). In the US, California is the most experienced water trading state, having experimented primarily with three types of water markets: local water markets, which exist in agricultural communities throughout the state; agriculture–urban transfers, which reallocate water from agricultural districts to cities in order to accommodate population growth; and a water bank enacted in the early 1990s in response to drought. However, the institutional and regulatory fragmentation described above has prevented any further water trading in the US. However, some states such as Oregon and Ohio have been expanding their water quality credit trading as an innovative way to achieve water quality goals more efficiently (EPA 2003). Trading is based on the fact that sources in a watershed can face very different costs to control the same pollutant. Trading programs allow facilities facing higher pollution control costs to meet their regulatory obligations by purchasing environmentally equivalent (or superior) pollution reductions from another source at lower cost, thus achieving the same water quality improvement at lower overall cost. While Australia has some experience in water quality credit trading in the Hunter salinity system, it is largely confined to that region and the concept is not covered in the NWI.

It would be hard to argue that there is anything other than a concerted shift towards the use of market-based instruments to achieve both environmental objectives, and equitable and sustainable supplies of water resources for consumptive uses. With the exception of the EU, which has confined its policy instruments to water pricing and educational campaigns, the case studies have embraced neo-classical economy theory to almost the same degree as Australia. It is worth pointing out, though, that the EU's decision not to introduce water trading, together with its complementary and extensive rural development and agri-environment policies, was a political decision based on the social and cultural values inherent in European society and is a marked rejection of the most hard-line economic rationalist approaches. Until Australia's NWI, and the water trading scheme upon which it rests, proves environmentally, economically and socially successful, we may well have to roll back some of our policies towards the more balanced European approach.

Indeed, the political commitments made in the NWI to water trading represent the most all-encompassing policy framework underscored by the principles of efficiency and economic rationality. The shift towards economic-based instruments for water management has yet to be fully implemented in any of the countries discussed. Indeed each, particularly Australia (owing to the full extent of its policy framework), faces significant challenges which could derail the whole exercise. In relation to full cost recovery water pricing, there is an underlying assumption that the environmental costs can be measured accurately. This has yet to be proven. In relation to water trading, it is evident that two unresolved challenges exist: the need to establish consistent water entitlements frameworks (yet to be done anywhere), and whether the political consequences that will arise when less-efficient sectors are forced out of the market can be managed with minimum fall-out.

Conclusion

This chapter aimed to draw on international developments in water policy and management for lessons that might assist NWI implementation. The importance of water resources as the 'bloodstream of the biosphere'(Ripl 2006) has been recognised across the case studies and has spawned reforms in policy in all but one (the US). It is evident that some common principles have emerged in water resource management, including:

- the development of (supra)national overarching policy frameworks
- the need to account for environmental water needs, although the legal basis for providing environmental water differs between jurisdictions
- the need to adopt an integrated holistic approach to water resource management with appropriate catchment-based institutions to implement policy
- the shift towards market-based instruments such as full cost recovery pricing and water trading
- the emphasis placed on public participation and community engagement in water policy implementation.

It is also evident that the significant challenges faced in implementing the NWI – identified in other chapters in this volume – are common to all jurisdictions. The ability of each country to actually achieve the policy commitments set out in the various frameworks remains to be seen, and will rely on further research.

There are, however, lessons for Australia's water reform agenda. On a positive note, Australia's experience with catchment management approaches and relatively long experience with catchment management authorities means that we are further along in knowing what financial, human and intellectual requirements they are likely to need in their now-significant role in water policy. Similarly, Australia's tentative steps towards developing nationally consistent datasets and water entitlements means that we are again further along the road than other countries, as some struggle to implement entitlements and licensing arrangements at all (i.e. South Africa). In this respect, we have a handle on the jurisdictional fragmentation that impedes greater reform in the US, for example.

In terms of areas to learn from, the difficulties inherent in defining environmentally sustainable levels of extraction are common to all jurisdictions. However, the EU's detailed and prescriptive approach in defining good ecological status – together with detail on the frequency and substance of monitoring and evaluation procedures – is an example to Australia. South Africa's four-tiered approach to environmental protection, while *prime facie* not as environmentally progressive as the NWI, may turn out to be more feasible as tensions between social, economic and environmental factors become more apparent. Finally, Australia's shift to water trading as the centrepiece of water policy should not be taken as either an end in itself, nor as a fait accompli.

References

Andreen W (2006). Developing a more holistic approach to water management in the United States. *Environmental Law Reporter* **36**, 10277–10289.

Arnold R (2005). Is wet growth smarter than smart growth? The fragmentation and integration of land use and water. *Environmental Law Reporter* **35**, 10152–10164.

CEC (2000). Directive 2000/60/EC of the European Parliament and of the Council of 23 October 2000 establishing a framework for community action in the field of water policy. *Official Journal of the European Communities* **L. 327/1**.

CEC (2003). Handbook on the implementation of EC environmental legislation. Available at http://www.europa.eu.int/comm/environment/enlarg/enlargement_en.htm. Accessed 14 November 2005.

Connell D, Dovers S & Grafton Q (2005). A critical analysis of the National Water Initiative. *Australasian Journal of Natural Resources Law and Policy* **10** (1), 81–107.

Daily GC (1997). *Nature's services: societal dependence on natural ecosystems*. Island Press: Washington DC.

de Coning C & Sherwill T (2004). An assessment of the water policy process in South Africa. WRC Report No TT232/04. Report to the Water Research Commission.

Dovers S (2005). *Environment and sustainability policy: creation, implementation, evaluation*. Federation Press: Sydney.

Dillion P, Benyon R, Cook P, Hatton T, Marvanek S & Gillooly J (2001). *Review of research on plantation forest water requirements in relation to groundwater resources in the southeast of South Australia*. Report No 99, Centre for Groundwater Studies.

DWAF (2004). National Water Resource Strategy, 1st edn (September). Available at http://www.dwaf.gov.za/Documents/Policies/NWRS/Sep2004/pdf/General.pdf.

Environment Canada. (2003). 1987 Federal Water Policy. Available at http://www.ec.gc.ca/WATER/en/info/pubs/fedpol/e_fedpol.htm. Accessed 24 November 2006.

EPA (2003). Water Quality Trading Policy. Available at http://www.epa.state.oh.us/dsw/WQ_trading/usepatradingpolicyfinal2003.pdf. Published 13 January 2003.

Falkenmark M (2003) *Water management and ecosystems: living with change*. TEC background paper no. 9, Global Water Partnership: Stockholm, Sweden.

Folke C (2006). Freshwater for resilience: a shift in thinking. *Philosophical Transactions* **358** (1440), 2027–2036.

Gunningham N (1999). Regulatory pluralism: designing policy mixes for environmental protection. *Land and Policy* **21** (1), 49–76.

Gunningham N & Grabosky P (1998). *Smart regulation: designing environmental policy*. Clarendon Press: Oxford.

Gunningham N & Sinclair D (2004). Curbing non-point source pollution: lessons from the Swan-Canning. *Environment and Planning Law Journal* **21**, 181–199.

Hansjugens B & Messner F (2002). Water pricing in the EU Water Frame Directive: Realizing full cost recovery and polluter-pays principles, *Wasser und Boden (Wasser Boden)* **54** (7–8), 66–69.

Hussey K (2006). Sustainable water management: comparative perspectives from Australia, Europe and the United States. *Environmental Law Reporter* **36** (4), 10259–10264.

Hussey K & Dovers S (2006). Trajectories in Australian water policy. *Journal of Contemporary Water Research and Education* **135** (December), 35–50.

Jones G, Hillman T, Kingsford R, McMahon T, Walker K, Arthington A, Whittington J & Cartwright S (2002). *Independent report of the expert reference panel on environmental flows and water quality requirements for the River Murray system*. Prepared for the Environmental Flows and Water Quality Objectives for the River Murray Project Board.Cooperative Research Centre for Freshwater Ecology: Canberra.

Keenan RJ, Parsons M, O'Loughlin E, Gerrand A, Beavis S, Gunawardana D, Gavran M & Bugg A (2004). *Plantations and water: a review*. Report prepared for Forest and Wood Products Research and Development Corporation. Bureau of Rural Sciences: Canberra.

Kuckshinrichs W & Schlör H (2005). Challenges for the German water sector: the EU Water Framework Directive, national infrastructure and demography. *Economic and Business Review* **7** (4), 291–309.

Lampinen R & Uusikyla P (1998). Implementation deficit: why member states do not comply with EU directives. *Scandinavian Political Studies* **21** (3), 231–251.

McLaughlin K, Osborne SP & Ferlie E (eds) (2002). *New public management: current trends and future prospects*. Routledge: London.

Mitchell WC & Simmon RT (1994). *Beyond politics: markets, welfare and the failure of bureaucracy.* Westview Press: Boulder, CO.

Ripl W (2006). Water: the bloodstream of the biosphere. *Philosophical Transactions* **358** (1440), 1921–1934.

Rockström J (2006). Water for food and nature in drought-prone tropics: vapour shift in rain-fed agriculture. *Philosophical Transactions* **358** (1440), 1997–2010.

Rockström J, Gordon L, Folke C, Falkenmark M & Engwall M (1999). Linkages among water vapor flows, food production, and terrestrial ecosystem services. *Conservation Ecology* **3** (2), 5. Available at http://www.consecol.org/vol3/iss2/art5/.

Smith SI (2001). *Water in Australia: resources and management.* Oxford University Press: Oxford.

UN (1999). *Freshwater. Global environmental outlook 2000,* ch. 2 Available at http://www.unep.org/geo2000/english/0046.htm.

Wild River S (2003). Local government. In *Managing Australia's environment.* (Eds S Dovers & S Wild River), 338–362. Federation Press: Sydney.

Endnotes

1 For instance, millions of wetlands have been destroyed; thousands of dams have disrupted aquatic ecosystems; land clearing and irrigation have in many instances led to higher levels of erosion and salinity; run-off from agricultural areas has led to streams contaminated with nutrients, minerals and pesticides; and industrial and municipal pollution have contaminated countless waters.

2 NWI para. 7. Other initiatives and policies which affect sustainable water management include the National Action Plan for Salinity and Water Quality, the National Water Quality Management Strategy and the Natural Heritage Trust. To the extent of any inconsistency between these and the NWI, the NWI prevails.

3 In the 1970–80s, consistent with the longer-standing US policy and legal debate, the term used in Australia regarding environmental water allocations was 'instream flows', including not only environmental but also recreational, cultural and aesthetic values. By the 1990s, the narrower construct of 'environmental flows' had become dominant. This shift bears closer examination.

4 National Groundwater Committee. Integrated groundwater: surface water management. Issue Paper 1.

5 'Environmental water manager' is defined as an expertise-based function with responsibility for management of environmental water and giving effect to the environmental water objectives of water plans. It may be a separate body or an existing basin, catchment or river manager.

6 The National Water Commission will comprise up to seven members, appointed for up to three years and eligible for reappointment subject to agreement. Commissioners will be appointed on the basis of experience in audit and evaluation, governance, resource economics, water resource management, freshwater ecology and hydrology. The Commonwealth can appoint four members (including the Chair); the states and territories may appoint three (NWI Schedule C).

Index